Ecology of
MARINE BIVALVES

An Ecosystem Approach

Marine Science Series

The CRC Marine Science Series provides publications that synthesize recent advances in Marine Science. Marine Science is at an exciting new threshold where new developments are providing fresh perspectives on how the biology of the ocean is integrated with its chemistry and physics.

CRC MARINE SCIENCE SERIES

Series Editors

Michael J. Kennish, Ph.D.
Peter L. Lutz, Ph.D.

Published Titles

Chemical Oceanography, 2nd Edition, Frank J. Millero
Ecology of Estuaries: Anthropogenic Effects, Michael J. Kennish
Ecology of Marine Invertebrate Larvae, Larry McEdward
Morphodynamics of Inner Continental Shelves, L. Donelson Wright
Ocean Pollution: Effects on Living Resources and Humans, Carl J. Sinderman
Physical Oceanographic Processes of the Great Barrier Reef, Eric Wolanski
The Physiology of Fishes, David H. Evans
Practical Handbook of Marine Science, 2nd Edition, Michael J. Kennish

---

Forthcoming Titles

Benthic Microbial Ecology, Paul F. Kemp
The Biology of Sea Turtles, Peter L. Lutz and John A. Musick
Chemosynthetic Communities, James M. Brooks and Charles R. Fisher
Coastal Ecosystem Processes, Daniel M. Alongi
Ecology of Marine Bivalves: An Ecosystem Approach, Richard F. Dame
Environmental Oceanography, 2nd Edition, Tom Beer
The Physiology of Fishes, 2nd Edition, David H. Evans
Practical Handbook of Estuarine and Marine Pollution, Michael J. Kennish
Seabed Instability, M. Shamim Rahman
Sediment Studies of River Mouths, Tidal Flats, and Coastal Lagoons, Doeke Eisma

Ecology of
MARINE BIVALVES
An Ecosystem Approach

Richard F. Dame, Ph.D.
Department of Marine Science
Coastal Carolina University
Conway, South Carolina

CRC

CRC Press
Boca Raton New York London Tokyo

Acquiring Editor: Paul Petralia
Cover Designer: Denise Craig
Marketing Manager, Direct Marketing: Arline Massey
Marketing Manager, Marketing Management: Susie Carlisle
Prepress: Kevin Luong
Project Editor: Helen Linna
Manufacturing Assistant: Sheri Schwartz

Library of Congress Cataloging-in-Publication Data

Dame, Richard F.
Ecology of marine bivalves: an ecosystem approach/Richard F. Dame
p. cm.— (CRC marine science series)
Includes bibliographical references and index.
ISBN 0-8493-8045-6
1. Bivalvia—Ecology. I. Title. II. Series.
QL430.6.D33 1996 96-11641
CIP

International Standard Book Number 0-8493-8045-6
Library of Congress Card Number 96-11641
Printed in the United States of America 1 2 3 4 5 6 7 8 9 0
Printed on acid-free paper

Preface

This book presents the first synthesis of marine bivalve ecology from an ecosystem or holistic view. It is designed to be a reference for bivalve researchers, coastal and fisheries resource managers, and students. The philosophy of the ecosystem approach is introduced and used as a template for the book. The concepts of the physical environment, organismic, and population processes are presented as foundations to understanding the later chapters on ecosystem processes. Ecosystem processes including grazing, system metabolism, and nutrient cycling in systems that are dominated by marine bivalves are concisely reviewed and a general conceptualization is developed. The utilization of ecosystem experiments in bivalve dominated systems is presented from the perspective of computer simulations, mesocosm studies and field manipulation experiments. The emerging field of ecosystem health is described from a bivalve viewpoint. Finally, the roles of marine bivalves as keystone species and ecosystem engineers are demonstrated with a discussion of interesting and pressing research issues.

Acknowledgments

The writing of this book would not have been possible without the limitless encouragement and support of my wife Meri. The untiring assistance of the Coastal Carolina University librarians is gratefully acknowledged. The following research colleagues were instrumental in discussing and reviewing various chapters: Dave Bushek of the USC Baruch Institute, Norbert Dankers of The Netherlands Institute for Nature Management, John Kraeuter of Rutgers University, Roger Newell of the University of Maryland, Theo Prins of The Netherlands Institute of Ecology, Hub Scholten of the University of Wageningen, Tim Schowalter of Oregon State University, Aad Smaal of the Ministry of Transport and Public Works of The Netherlands, and Ray Seed of the University of Wales. Time for writing the book was provided by a Palmetto Professorship to the author and research days from the National Science Foundation. The author was supported by research funds from NOAA and NSF during the production of the text.

The Author

Richard F. Dame, Ph.D., is Professor of Marine Science at Coastal Carolina University in South Carolina.

Dr. Dame received his B.S. degree from the College of Charleston, Charleston, South Carolina in 1964. He obtained his M.A. degree from the University of North Carolina in 1967 and his Ph.D. degree from the University of South Carolina in 1971. He was appointed Assistant Professor of Marine Science at the then USC — Coastal Carolina College and Research Associate of the Belle W. Baruch Institute for Marine Biology and Coastal Research where he still remains. He was Chair of the Marine Science Department from 1981–1991. From 1992–94, he was the Ecosystem Studies Program Director at the National Science Foundation.

Dr. Dame is a member of numerous scientific societies and is currently the President of the Southeastern Estuarine Research Society. He was named the Distinguished Alumnus of the College of Charleston in 1989 and also named the first Palmetto Professor in South Carolina.

He has received numerous research grants and awards from the National Science Foundation, National Oceanographic and Atmospheric Administration, Sea Grant, North Atlantic Treaty Organization and U.S. Environmental Protection Agency. He has published over 60 peer reviewed articles and edited 2 books. His current research is focused on the role of oysters in estuarine ecosystems.

Contents

Introduction

Supporting Processes

Chapter 4 Population Processes

Ecosystem Processes

Chapter 5 Grazing

Chapter 8 Bivalves as Components of Ecosystem Health

Synthesis and Conclusions

Chapter 9 Conclusions

INTRODUCTION

1 Ecosystem Perspective and Bivalve Molluscs

EVOLUTIONARY HISTORY

The first bivalve molluscs appeared in the sea during the Cambrian period about 620 M years ago, or before organisms had invaded the land (Figure 1). Because their shells were made of calcium carbonate, bivalves are well represented in the fossil record and often used as paleontological markers. During the Paleozoic (620 to 230 M years ago), the bivalve molluscs were overshadowed by an unrelated but similarly structured group, the Brachiopods or lamp shells. Both groups were filter feeders and removed phytoplankton from the water column by means of a ciliary feeding mechanism. Gradually the bivalve molluscs increased in importance so that by the Mesozoic era they were dominant and remain so today (Yonge 1960). The basic bivalve plan has been successful for many hundreds of millions of years. Due to the similarities between some fossil forms and extant species, it has often been said that an oyster of 200 M years ago would be perfectly familiar if served in a restaurant today (Yonge 1960).

The bivalve molluscs are thought to have originated in warm shallow euhaline coastal waters and gradually invaded estuaries and brackish systems, as well as all the reaches of the world ocean. Because the adult forms of the majority of these animals are benthic or bottom dwelling, many different evolutionary adaptations to the benthic habitat have occurred. The most common life styles are: buried within burrows in unconsolidated soft sediments; attached by byssal threads to pebbles or more consolidated sediments; cemented to shells or rocks; and as semimobile members of the epibenthos. Today these adaptations are exemplified by clams, mussels, oysters, and scallops, respectively. In nature there are often gradients of sediments from muddy unconsolidated materials to hard substrates that reflect a water dynamics-energy gradient from low energy to high energy environments, with bivalves zoned accordingly (Morton 1992).

As dense associations of bivalves are likely to have the greatest influence on their ecosystems, when and where did reefs and beds of these organisms appear? Primitive and long extinct bivalves, the rudists, appeared in the Paleozoic (Figure 2). They were gregarious in nature and were often found in association with corals and other reef-forming organisms (Stenzel 1971). Later, the fossil gryph-shaped oysters of the Miocene are often found associated with corals and other euhaline reef-building forms. This suggests that the advantages of reef building were exploited by bivalves at an early stage in their evolution.

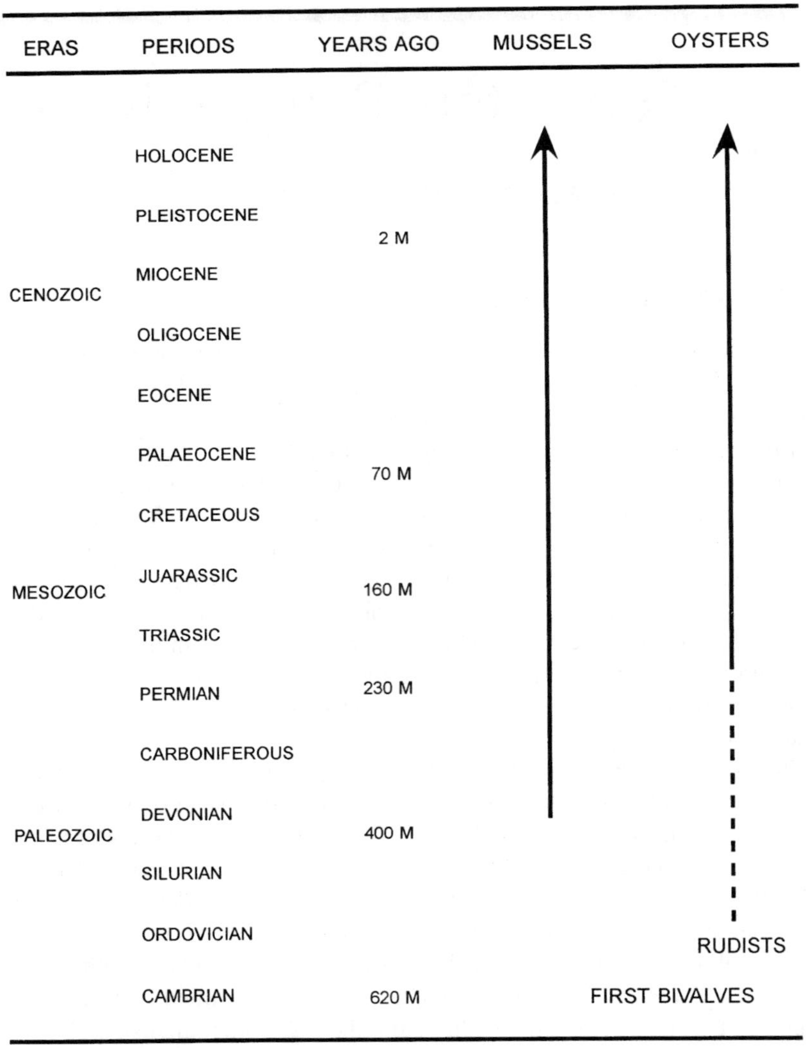

FIGURE 1. Evolutionary history of bivalve molluscs. (Data from Yonge 1960, Stenzel 1971, and Morton 1992.)

FIGURE 2. Rudist reef. (Adapted from Kauffman and Sohl 1974.)

The largest reef building bivalve known was *Crassostrea gryphoides,* the giant oyster, also known as *C. gigantissima*, in North America. These bivalves with a reported height of 60 cm, weight of 4.5 kg, and shell thickness of 15 cm grew rapidly in the warm shallow brackish waters of the Miocene (Stenzel 1971). Today, bivalve reefs are only formed by brackish water species such as *Crassostrea* and *Mytilus.* In contrast to coral reefs, these reefs are found mainly in the temperate zone (Hughes 1991). In most cases, as the bivalve shell framework accumulates, it eventually becomes unstable and collapses. However, in the southeastern U.S.A., oyster reefs dominated by *C. virginica* can reach extensive size and persist for long periods of time (Hughes 1991).

The first clear attempt relating living bivalves to their associated organisms and physical environment was by Möbius (1880). In his studies of oyster beds along the Schleswig-Holstein shore of Germany and the estuaries of England, he noticed that the organisms found there had numerous things in common. First, the same species composition of bivalves, crabs, barnacles, bryozoans, starfish, sea urchins, worms, fish, and algae were present. Möbius speculated that every oyster bed is a collection of organisms that find everything necessary for their growth and reproduction. Second, these organisms were normally found with similar environmental qualities of food, salinity, temperature, and substrate. He coined the word "bioceonosis" to describe the community of living organisms that are mutually limited by the environment. Möbius also observed that when the external environment changed, so did the composition of the bed. This effective binding of the community of organisms to the vagaries of the environment is a strong predilection toward what today we think of as ecosystems.

ECOSYSTEMS

Ecosystems are organized assemblages of plants and animals in complex interaction with each other and their physical environment. While individual species are governed by genetic and physiological processes, their existence in nature is conditioned by the total environment and in turn influences the environment of which they are a part. Because of such complex interdependencies, understanding ecosystems requires studies that holistically elucidate principles of total system function. For the purposes of this book, I will use Odum's (1983) definition for ecosystems:

> an ecosystem is any unit that includes all organisms that function together in a given area interacting with the physical environment so that the flow of energy leads to clearly defined biotic structure and cycling of materials between living and non-living parts.

Within the ecosystem approach, the structural components of biomass (including skeleton), hierarchy, food webs, flow networks, and biodiversity are important in determining the role of bivalves in ecosystems. Energy flow and material cycling processes, e.g., feeding, respiration, growth, excretion, and predation, as well as control by feedback loops, all help describe the functional roles bivalves play within marine and coastal ecosystems (Figure 3). Quite often the magnitude of functional processes is closely coupled to the biological and physical aspects of structure.

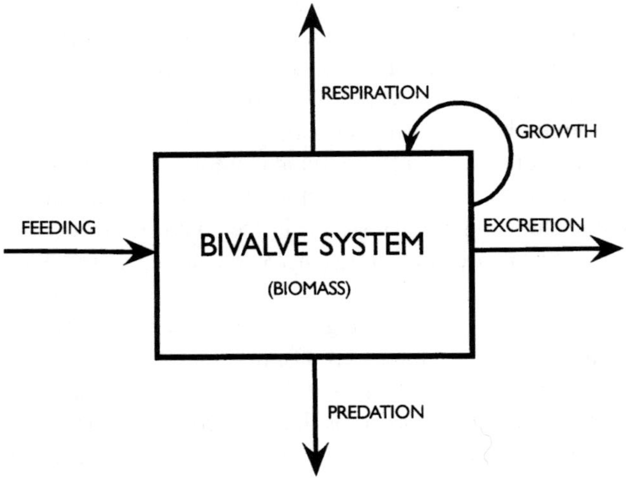

FIGURE 3. Functional processes of bivalve systems.

Structural Attributes

The most obvious and tangible contribution of bivalves to the structure of ecosystems is the biomass, skeleton and burrows of the animals themselves. The shells of bivalves not only provide a platform for these animals, but when taken with the soft body, amount to major components within some ecosystems. These structures also form a system matrix within which habitats are generated for other organisms. This matrix allows an increase in species diversity and leads to increased system complexity and redundancy (Hughes 1991).

The physical environment is also a component of ecosystem structure. Not only can abiotic factors limit the abundance and extent of the biota, but the living components can influence the physical environment. Because bivalve shells are rigid, the flow of water around them can be both microscopically and macroscopically altered. In dense associations of bivalves in the form of reefs or beds, the accumulated empty and living shells of bivalves can markedly alter the physical dynamics of water flow over these structures at even greater physical scales, i.e., creek or channel. Generally, the orientation of the individual bivalves within the bed tends to enhance water circulation and turbulence over the system and in turn enhances the functional performance of the molluscs (Hughes 1991).

System Hierarchy

Recently, the development and application of the ideas of levels of spatial and temporal organization, now known as hierarchy theory, have blossomed in ecosystem ecology (O'Neill et al. 1986). These authors note that the relationships between the organisms in a system and the functions of the system are not always clear. They contend that hierarchy theory provides a framework for understanding this relationship without reducing one area of emphasis to the other. Two separate hierarchies, the population–community and the process–functional approaches, have emerged, which address ecosystems differently. The former is based on like taxonomic groupings, while the latter is predicated on similar ecological processes. Essentially, the two different fields view the natural world from different observational perspectives.

The population–community approach is tangible and envisions a sequence of ORGANISM–POPULATION–COMMUNITY (Figure 4) and this hierarchy generally views the physical environment as a backdrop to the biological processes (O'Neill et al. 1986). The process–functional school places organisms of like functional roles together (Figure 4) with the functional group at the base and the biosphere at the top (O'Neill et al. 1986). This hierarchy is less tangible than the population–community hierarchy, but the physical environment is interwoven conceptually.

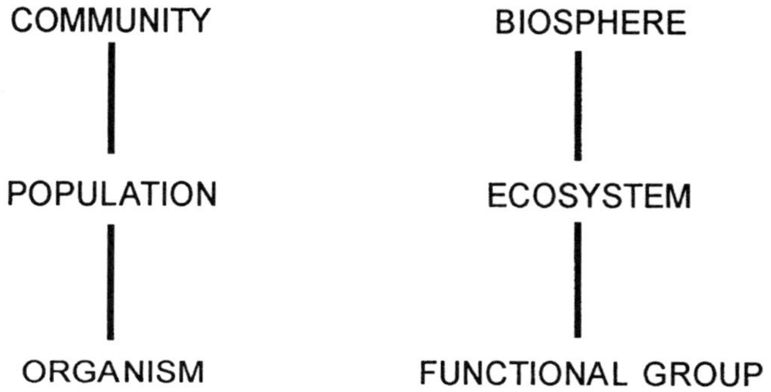

FIGURE 4. Simple hierarchies in ecosystems. (Adapted from O'Neill et al. 1986.)

There is no immediately recognizable one-to-one or even many-to-one correspondence between the two hierarchies (O'Neill et al. 1986). Generally, components of neither of the two hierarchies can be reduced to each other in any simple way. It is also well known from observations of functional redundancy that many species can perform essentially the same ecosystem task. It is also probable that changes in the biodiversity of a system can in some cases influence ecosystem function. Further, many ecosystem functions, i.e., nutrient cycling, involve components that are abiotic (e.g., sediments) or combine biotic and abiotic elements. Obviously there is no easy correspondence of such components with species or assemblages of species.

Often the population–community and process–functional hierarchies are difficult to rectify with each other. However, in some ecosystems, a single species or a group of closely related species may dominate a process or functional component. In such cases, there is a direct correspondence between the population–community and process–functional hierarchies (Figure 5). A classical example is the former dominance of a single population of large ungulate herbivores, bison, in the extensive grassland ecosystems of North America. Likewise, in estuaries, extensive populations of mussels or oysters will dominate the herbivore component of these shallow water ecosystems and provide a potentially useful correspondence between the two hierarchies (Dame 1993).

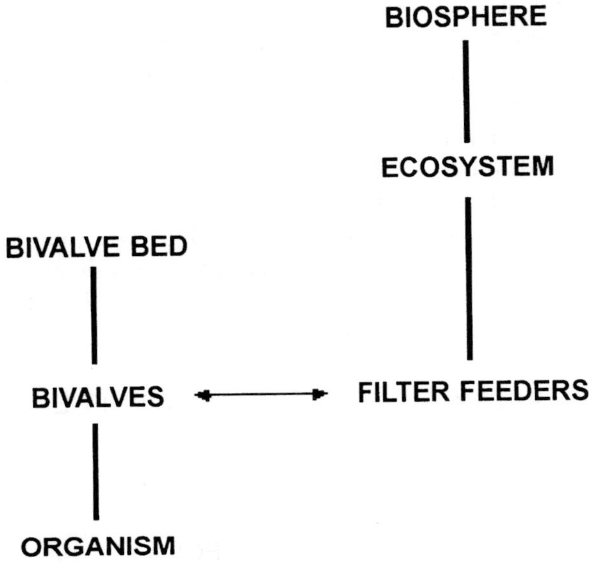

FIGURE 5. Correspondences between population–community and process–functional hierarchies in bivalve-dominated systems.

Food Webs

Food energy or matter is transferred through the organisms that make up an ecosystem, and the interlocking relationships between the various biological components provide structure to the ecosystem and are called a food web. Description of the components and the relationships between them is a long-held aspect of ecology. Only recently have issues arisen concerning the influence or control, both quantitative and qualitative, of particular components on their food web. Initially these questions were solely from the population–community perspective, but recently a broader ecosystem perspective appears to be emerging.

The paradigm of bottom-up control of food chains (Figure 6), with nutrients and light of major importance to primary producers and resulting material flows up the food chain to affect the remaining trophic levels, is well founded historically (Lindeman 1942). The top-down (cascading) or food chain dynamics theory was first proposed by Hairston et al. (1960) and was recently more fully developed by Fretwell (1987) and Oksanen (1988). This theory predicts that the top predator controls the next lower trophic level, resulting in alternating magnitude (negative correlation) of biomass and turnover patterns for adjacent trophic levels. Many of the speculations concerning bivalve filter feeders controlling phytoplankton abundances in the water column subtly fall into this category.

The mutualism–cybernetics theory of Odum and Biever (1984) suggests that plant–herbivore interactions are typically donor controlled or mutualistic (positive correlation), as herbivores often stimulate the production of the plant species that they feed on. The latter theory is favored by the process–functional group, but attacked by the population–community ecologists because it implies community

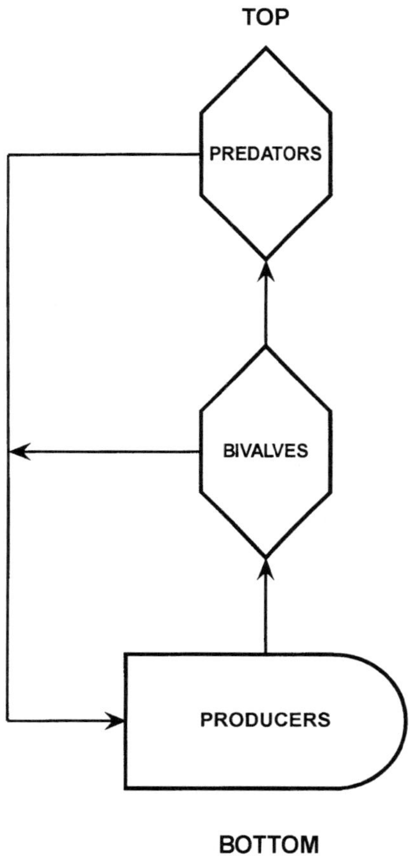

FIGURE 6. Food chain controls using energy symbols.

selection or evolution above the species level. Because of these criticisms, Vadas (1989) has proposed another theory, the donor control–indirect effects theory. In this idea, species interact at ecological time scales with indirect effects caused by nutrient cycling, competition, and abiotic fluctuations leading to donor-controlled predator–prey interactions. More recently, Hunter and Price (1992) and Menge (1992) have argued in a similar vein that bottom-up and top-down controls operate simultaneously in most systems, and it is the net pattern that we observe in nature.

I know of only two studies to date that address the food web control issue in ecosystems dominated by bivalves. In a mainly speculative vein, Menge (1992) has argued that both top-down and bottom-up controls operate simultaneously in shallow water systems coupled to the rocky intertidal zone along the Oregon coast. In these systems, the mussel, *Mytilus*, is a dominant component, and different scenarios are envisioned, depending on whether or not one system, Boiler Bay, is driven by benthic plants (Figure 7) or the other, Strawberry Hill, is controlled by phytoplankton (Figure 8).

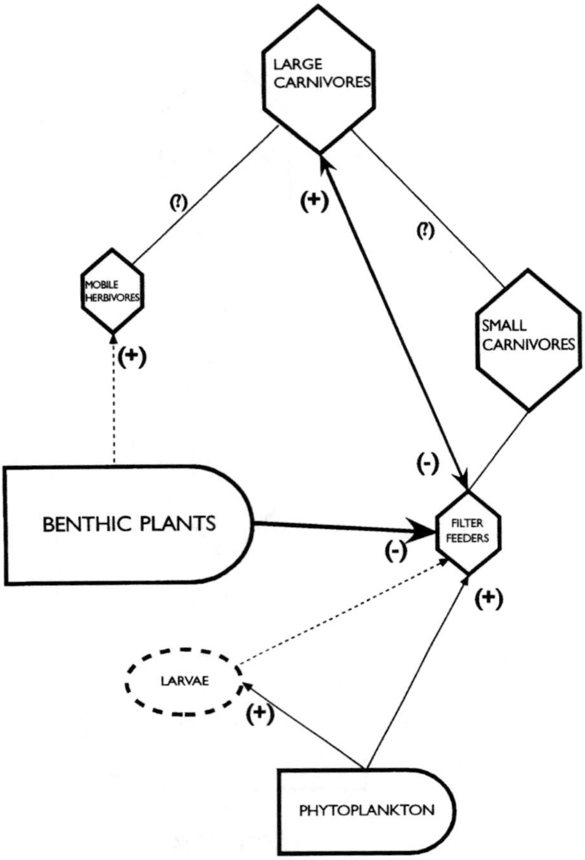

FIGURE 7. Hypothetical interaction food web for Boiler Bay, OR, with size of compartment indicating abundance and thickness of line indicating strength of relationship. (Adapted from Menge 1992.)

In a series of intense ecosystem level studies, Dame et al. (1989, 1991) investigated the role of oysters and mussels on material processing in estuaries. Utilizing a preliminary field experiment, Dame and Libes (1993) showed that the removal of oysters from a series of tidal creeks did not influence phytoplankton abundance as measured by chlorophyll *a* or water column primary production, but did significantly reduce the concentrations of dissolved inorganic nutrients. Thus these results within small creeks and on short time scales, imply a potential dominance of both bottom-up and top-down controls within these systems.

Networks

Food chains and webs are simple conceptualizations of natural ecological systems. In order to develop deeper understandings of the dynamics of ecosystems, ecologists

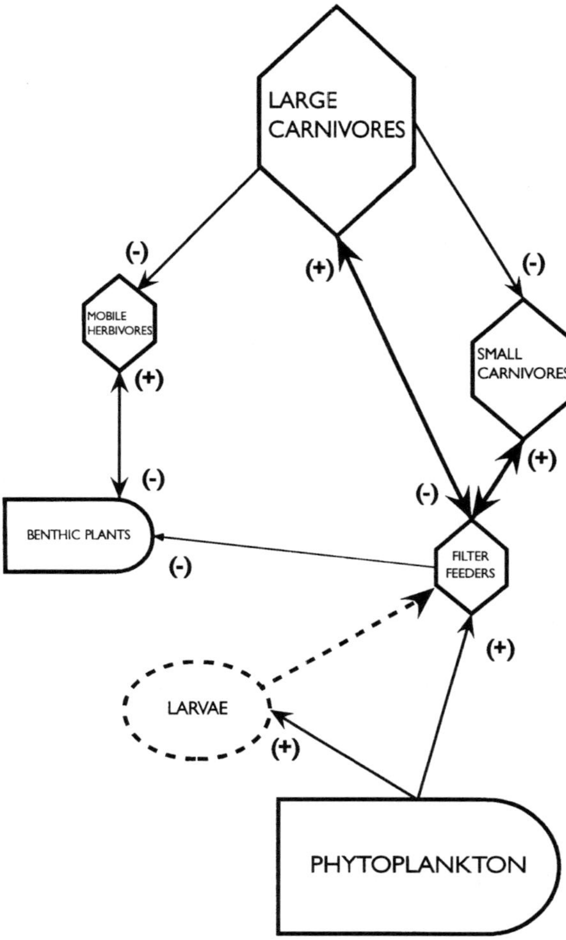

FIGURE 8. Hypothetical interaction food web for Strawberry Hill, Oregon as described in Figure 7. (Adapted from Menge 1992.)

have constructed much more complicated models. Some models have been designed for computer simulation and utilize systems of differential equations, while others are created for analysis. From a process–functional point of view, the measurement and depiction of material and energy flow between the various ecosystem components can show the efficiency with which material is transferred, assimilated, and dissipated. Such conceptualizations (see Chapter 7) convey considerable information on the structure of the ecosystem (Baird et al. 1991). Collectively known as network analysis, input–output, trophic, and cycle analysis have been applied in an engineering sense to a number of bivalve-dominated systems, an oyster reef in South Carolina (Dame and Patten 1981), Chesapeake Bay (Baird and Ulanowicz 1993), and several less-studied systems (Baird et al. 1991).

Biodiversity

Diversity is inherent in all levels of biological organization, from the molecules that carry genetic information, to species where natural selection takes place, to ecosystems where organisms interact with each other and their environment. From this perspective, biodiversity is a fundamental property of the structure and function of ecosystems (Ray 1991). In a process–functional view, if the diversity of organisms changes within a system, it is not clear what will happen to the functional attributes of the system in question. Although a fair amount is known about the diversity of bivalve mollusc species through space and time, molecular diversity in bivalves is only known in a few commercially important species, and the role of bivalves in biodiversity and ecosystem function is relatively unknown.

Functional Attributes

At the ecosystem level, the most important functional processes deal with the flow of energy and cycling of materials between and within systems. Traditionally, groups of organisms with similar functions, processes, feeding, or trophic levels have been grouped together into functional components. The most common functional groups represented by bivalves are filter feeders, deposit feeders, consumers, and herbivores. Each component has inflows and outflows that are represented by specific organismic level processes, e.g., feeding and excretion (Figure 3). As energy and matter are processed and transformed by the organisms, work is done and degraded energy or waste heat is lost from the system. This latter process is known as respiration at both the organismic and system levels. Production or growth is the storage of energy and matter by a given system component. These processes may result in the increase in size of individuals or populations of animals over a unit of time. Predation is the removal of energy and materials by one component feeding on another, usually animals eating animals. The activity of herbivores eating plants or grazing may also be termed predation.

Nutrient cycling is the process of specific types of nutrients or molecules moving from component to component within the ecosystem. Since matter is finite, atoms are used and reused within the ecosystem and the rates at which this reuse takes place is an exceedingly important process in ecosystem ecology. By determining nutrient cycling pattern and rates, ecosystems can be compared and important components in the cycling of a particular nutrient within ecosystems can be identified. In some ecosystems, bivalves are thought to increase the rate of nutrient cycling because of their high abundances and high metabolic rates (Table 1).

System Types

A variety of bivalve molluscs are prominent members of many diverse marine systems (Figure 9). These animals have reached their greatest abundances in shallow coastal waters and estuaries. In these systems, phytoplankton is consumed from tidal waters sweeping over dense beds of bivalves, mainly clams, cockles, mussels, oysters, and scallops. The dominant bivalves in these systems are adapted to changing temperatures and salinities and some are adapted to exposure and wave action in

TABLE 1
Spatial and Temporal Scales for Processes Involving Bivalve Molluscs

Process	Spatial Scale	Duration	Frequency
Physiological	cm–m	Minutes	None–seasonal
Predation	1–10 cm	Minutes–hours	Weeks–months
Associations	1–10 cm	Days	Months–years
Disease	cm–km	Days–weeks	
Temperature–salinity	1–100 km	Days	Months–years
Storms	1–100 km	Days	Months–decades
Epidemics	10–1000 km	Years	Decades–centuries
Climate change	Global	10–100,000 years	10,000–100,000 years

the intertidal zone. Three different life styles are prevalent: gregarious permanently attached forms, oysters; gregarious byssally attached and partially mobile forms, mussels; and bivalves that burrow into semi- or unconsolidated sediments, clams. Some bivalves have been successful in the extreme environments of the rocky intertidal zone and the sandy beach. In the former, rapid growth and a protective shell are major attributes, while in the latter, rapid burrowing in shifting sediments is important. In the low productivity zones of coral reefs and the deep sea vents, endosymbiotic relationships allow bivalves to succeed. Finally, shipworms succeed with specialized functions in eating and decomposing woody debris washed into the sea from the land.

Remaining Contents

Having presented a broad ecological setting, the remainder of this book focuses on the specific attributes of bivalve molluscs that allow them to play an important role in marine ecosystems. This presentation does not address bivalve genetics (see Gosling 1992 for an introduction) or aquaculture (see Menzel 1991 for a review) specifically, but there are many implicit references to the latter. Using the classical ecosystem approach, the interactions of bivalves with the physical environment, particularly temperature, salinity, and water motion are reviewed first. These interactions can have a major influence on rate processes or functions at the organismic level. Thus, the roles of filtration, respiration, growth, and reproduction are examined from an ecosystem perspective. Population and community scale interactions are addressed with particular attention given to growth, predation, and refugia in the well-studied mussel and oyster systems. The ecosystem process of bivalves grazing on primary producers is considered, and the potential for bivalves to control primary production is carefully examined. Intimately related to primary production, metabolism and nutrient cycling are described for several systems to define the role of bivalves as potential high speed processors of reactive nutrients. Because the sum of the preceding processes is exceedingly complex, some of the major conceptual and computer simulation models in which bivalves play a central role are reviewed. As it is difficult to test hypotheses only with models, ecosystem scale experiments

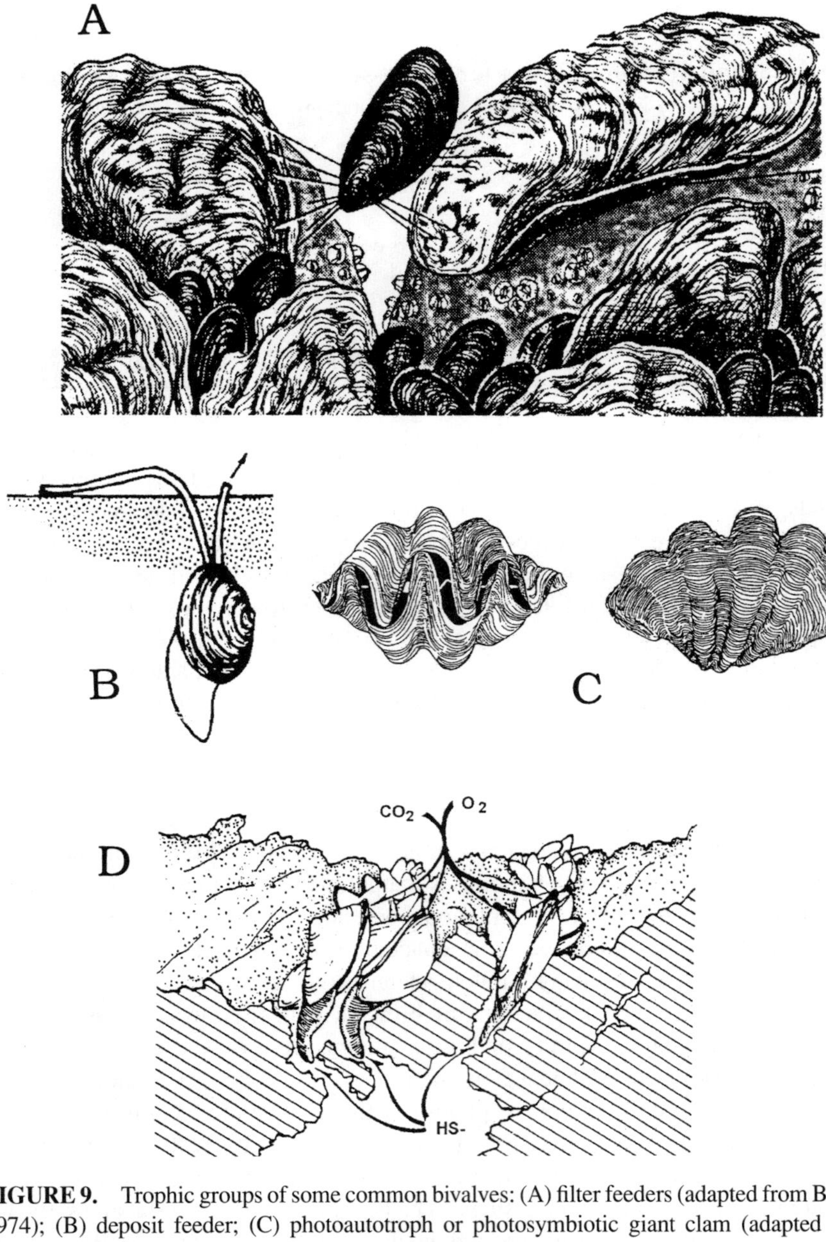

FIGURE 9. Trophic groups of some common bivalves: (A) filter feeders (adapted from Barnes 1974); (B) deposit feeder; (C) photoautotroph or photosymbiotic giant clam (adapted from Lucus 1988); (D) chemoautrotroph or chemosymbiotic mussels (adapted from Fisher 1990).

are described with their inherent strengths and weaknesses. Recognizing that marine bivalves are proven monitors and indicators of ecosystem stress, their role in ecosystem health is summarized. In the final chapter, the various aspects of bivalves in ecosystems are drawn together and the areas of most pressing research need are restated.

REFERENCES

Baird, D., McGlade, J.M., and Ulanowicz, R.E. 1991. The comparative ecology of six marine ecosystems, *Phil. Trans. R. Soc. Lond. Ser. B*, 333, 15–29.

Baird, D. and Ulanowicz, R.E. 1989. The seasonal dynamics of the Chesapeake Bay, *Ecol. Monogr.*, 59, 329–364.

Barnes, R.D. 1974. *Invertebrate Zoology*, 3rd ed., Saunders, Philadelphia, 870 pp.

Dame, R.F. 1993. The role of bivalve filter feeder material fluxes in estuarine ecosystems, in Dame, R.F., Ed., *Bivalve Filter Feeders and Coastal and Estuarine Ecosystem Processes*, Springer-Verlag, Heidelburg, 245–269.

Dame, R.F., Dankers, N., Prins, T., Jongsma, H., and Smaal, A. 1991. The influence of mussel beds on nutrients in the Western Wadden Sea and Eastern Scheldt estuaries, *Estuaries*, 14, 130–138.

Dame, R.F. and Libes, S. 1993. Oyster reefs and nutrient retention in tidal creeks, *J. Exp. Mar. Biol. Ecol.*, 171, 251–258.

Dame, R.F. and Patten, B.C. 1981. The analysis of energy flows in an intertidal oyster reef, *Mar. Ecol. Prog. Ser.*, 5, 115–124.

Dame, R.F., Spurrier, J.D., and Wolaver, T.G. 1989. Carbon, nitrogen, and phosphorus processing by an oyster reef, *Mar. Ecol. Prog. Ser.*, 54, 249–256.

Fisher, C.R. 1990. Chemoautotrophic and methanotrophic symbioses in marine invertebrates, *Rev. Aquat. Sci.*, 2, 399–436.

Fretwell, S.C. 1987. Food chain dynamics: the central theory of ecology?, *Oikos*, 50, 291–301.

Gosling, E.M. 1992. Genetics of *Mytilus*, in Gosling, E., Ed., *The Mussel Mytilus, Ecology, Physiology, Genetics and Culture*, Elsevier, New York, 309–382.

Hairston, N.G., Smith, F.E., and Slobodkin, L.B. 1960. Community structure, population control, and competition, *Am. Nat.*, 94, 421–424.

Hughes, R.N., 1991. Reefs, in Barnes, R.S.K. and Mann, K.H., Eds., *Fundamentals of Aquatic Ecology*, Blackwell Scientific, Oxford, 213–229.

Hunter, M.D. and Price, P.W. 1992. Playing chutes and ladders: heterogeneity and the relative roles of bottom-up and top-down forces in natural communities, *Ecology*, 73, 724–732.

Kauffman, E.G. and Sohl, N.F. 1974. Structure and evolution of Antillean Cretaceous rudist frameworks, *Verh. Naturforsch. Ges.*, 84, 399–467.

Lindeman, R.L. 1942. The trophic-dynamic aspect of ecology, *Ecology*, 23, 399–418.

Lucus, J.S. 1988. Giant clams: description, distribution and life history, in Copland, J.W. and Lucas, J.S., Eds., *Giant Clams in Asia and the Pacific*, Australian Center for International Agricultural Research, Canberra, 21–32.

Menge, B. 1992. Community regulation: under what conditions are bottom-up factors important on rocky shores?, *Ecology*, 73, 755–765.

Menzel, W. 1991. *Estuarine and Marine Bivalve Mollusk Culture*, CRC Press, Boca Raton, FL, 362 pp.

Möbius, K. 1880. Die Auster und die Austerwirthschaft, *Rep. U.S. Comm. Fish.*, 683–751.

Morton, B. 1992. The evolution and success of the heteromyarian form in the Mytiloida, in Gosling, E., Ed., *The Mussel Mytilus, Ecology, Physiology, Genetics and Culture*, Elsevier, Amsterdam, 21–52.

Odum, E.P. 1983. *Basic Ecology*, Saunders, Philadelphia, 603 pp.

Odum, E.P. and Biever, L.J. 1984. Resource quality, mutualism, and energy partitioning in food chains, *Am. Nat.*, 124, 45–52.

Oksanen, L. 1988. Ecosystem organization, mutualism and cybernetics or plain Darwinian struggle for existence?, *Am. Nat.*, 131, 424–444.

O'Neill, R.V., DeAngelis, D.L., Waide, J.B., and Allen, T.F.H. 1986. *A Hierarchical Concept of Ecosystems*, Princeton University Press, Princeton, NJ, 253 pp.

Ray, G.C. 1991. Coastal-zone biodiversity patterns, *Bioscience*, 41, 490–498.

Stenzel, H.B., 1971. Oysters, in Moore, R.C., Ed., *Treatise on Invertebrate Paleontology,* Part N, *Mollusca 6, Bivalvia,* Geological Society of America, Lawrence, KS, N593–1051.

Vadas, R.L. 1989. Food web patterns in ecosystems, a reply to Fretwell and Oksanen, *Oikos*, 56, 339–343.

Yonge, C.M. 1960. *Oysters*, Collins, London, 209 pp.

SUPPORTING PROCESSES

2 Physical–Environmental Interactions

INTRODUCTION

From an ecosystem perspective, the biotic and abiotic components of a given environment are intimately bound together to control and generate the ecological system. In many instances, there is a two-way interaction: the physical environment may control living components and the biotic environment may control abiotic components.

The existence of an organism in a given environment implies that the organism has succeeded or survived in the presence of all the abiotic and biotic influences that impinge upon it. Any condition or factor that approaches or exceeds the organism's tolerance for that factor is defined as a limiting factor or condition (Odum 1983). Studies on the lower limits of an organism's existence to physical factors were initiated by Liebig (1847) and later formalized into Liebig's "Law of the Minimum." From Liebig's studies, it was found that the law was strictly applicable only to systems in steady-state, i.e., systems where inflow balances outflow of energy and materials. In addition, Liebig found that factors or conditions can influence each other, and care must be taken to recognize this potential interaction.

Organisms have a life zone, often called the "biokinetic zone" (Vernberg and Vernberg 1972), where there are both lower (minima) and upper limits for a given factor. This concept was first developed by Shelford (1913) and is generally termed Shelford's "law of tolerance." Organisms with broad ranges of tolerance to a given condition are termed "eury-," while those with narrow biokinetic zones are called "steno-" (Figure 10).

In the remainder of this chapter, the major abiotic environmental factors that influence bivalve molluscs will be reviewed. The focus will be on the so called "operationally significant" factors that play a major role on bivalves during their life cycles.

TEMPERATURE

Temperature is a measure of the degree of random motion or kinetic energy in molecules of matter. Temperature not only limits the spatial distribution of bivalves, but also is a major controlling factor in many physiological rate processes, e.g., feeding and growth, that play important roles in many ecosystem level processes. Because bivalve molluscs produce only small amounts of heat during metabolism, they are termed "poikilotherms," or animals whose body temperature conforms to

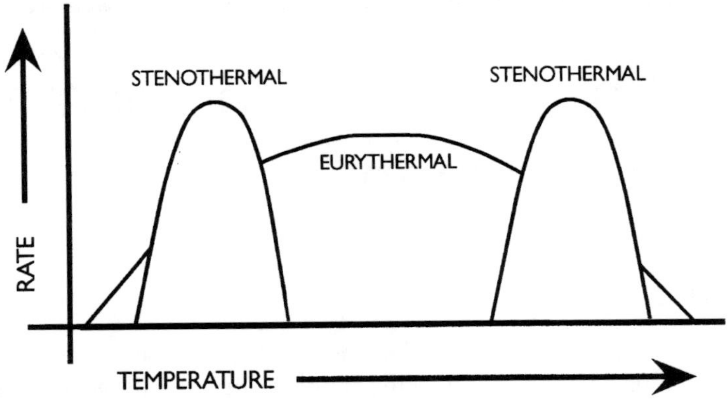

FIGURE 10. The generalized influence of temperature on rate processes in stenothermal and eurythermal bivalves. (Adapted from Shelford 1913.)

that of the environment. The majority of marine bivalves live within a temperature range from near –3°C to +44°C (Vernberg and Vernberg 1972). At the extremes, for example, intertidal oysters, *Crassostrea*, may attain body temperatures of 49°C in warm subtropical and tropical climates (Galtsoff 1964), and *Mytilus* in Canada have been shown to survive temperatures as low as –16°C (Bourget 1983), while coastal and estuarine bivalves tend to be eurythermal and oceanic forms are typically stenothermal in tolerance (Figure 10).

Incident light input per unit area of surface decreases from the equator to the poles due to the curvature of the earth. This decrease in input energy results in a latitudinal temperature gradient of decreasing temperature with increasing latitude. The gradient in temperature may result in broad geographical zonation of bivalve species. For example, along the Atlantic coast of North America *Mytilus edulis* is common to Cape Hatteras at the southern limit of Virginia and absent south of that location. It is generally believed that the warm water, northward-moving currents (the Gulf Stream), so typical of the western boundaries of oceans, provide a temperature barrier to the distribution and survival of *Mytilus* larvae southward. In contrast, *C. virginica* is a more eurythermal species and is found over a broader latitudinal range both north and south of Cape Hatteras. It is important to note, as with many other species, that it is the tolerance of the mobile larval stage or reproductive period that is most sensitive to temperature.

In addition to geographical distribution, physiological rates may vary over a latitudinal gradient. Butler (1953) has shown that for *C. virginica* growth rate of soft body is greatest near the central portion of this species' north–south range, or about the location of Chesapeake Bay (Figure 11). North and south of that point growth rate declines. Along the coast of the Gulf of Mexico, Butler speculates that the oysters are near the limit of their geographical distribution. Excluding nongrowth months, Butler's observations suggest that oysters from the warmest areas produce the least amount of biomass per unit time, and all other regions are considered

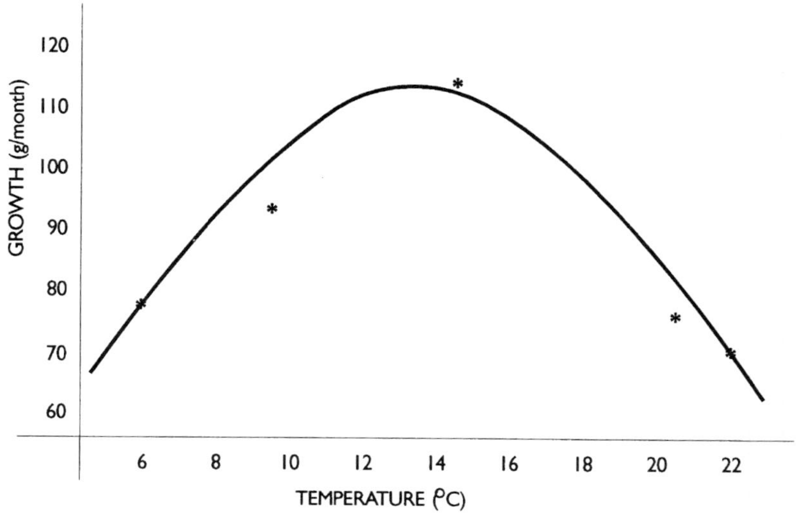

FIGURE 11. The influence of environmental temperature on the growth rate of *Crassostrea virginica* along a latitudinal gradient. (Data from Butler 1953.)

intermediate in growth rate. His data appear to fit the general form of the curve in Figure 10.

Butler's findings imply a more general application of perturbation theory and the subsidy-stress gradient approach (Odum et al. 1979). In this application, the input perturbation, temperature, is an energy source and a convex performance curve may simulate the system response (Figure 12). For the oyster case, as temperature increases, growth rate may increase and then gradually decrease. In addition, the relative variability of observed growth rate will increase as temperature increases. If the perturbation is destructive, as high temperatures are, then performance may be reduced to the point where the species and the community are replaced with more tolerant forms.

In addition to latitudinal temperature gradients, subtropical, temperate, boreal, and arctic coastal water habitats experience seasonal cycles in temperature. Seasonality is caused by the inclination of the earth as it spins on its axis in relation to the sun. For bivalves, seasonality reaches its greatest extent in temperate and boreal shallow tidal waters where land–sea interactions have their greatest influence on temperature.

Because organisms experience quite different temperature regimes depending on the time of year, they are essentially thermally acclimated to the different conditions in different seasons. Thermal acclimation is defined (Crisp and Ritz 1967) as any nongenetic adjustment by an organism in direct response to a change in a single factor in the environment. Many physiological rates, i.e., growth, respiration, feeding, excretion, etc., respond to thermal acclimation. In the case of intertidal oysters (Dame 1972), *C. virginica*, clearly respiration is significantly different at different acclimation temperatures (Figure 13). Thus, an organism can be seasonally

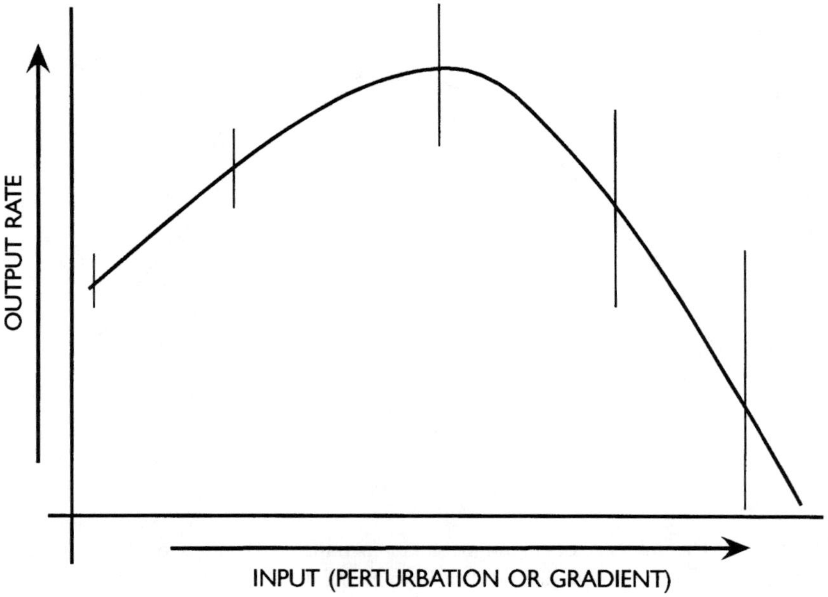

FIGURE 12. The general form of the subsidy-stress gradient approach and its similarity to oyster growth across a latitudinal gradient. (Modified from Odum et al. 1978.)

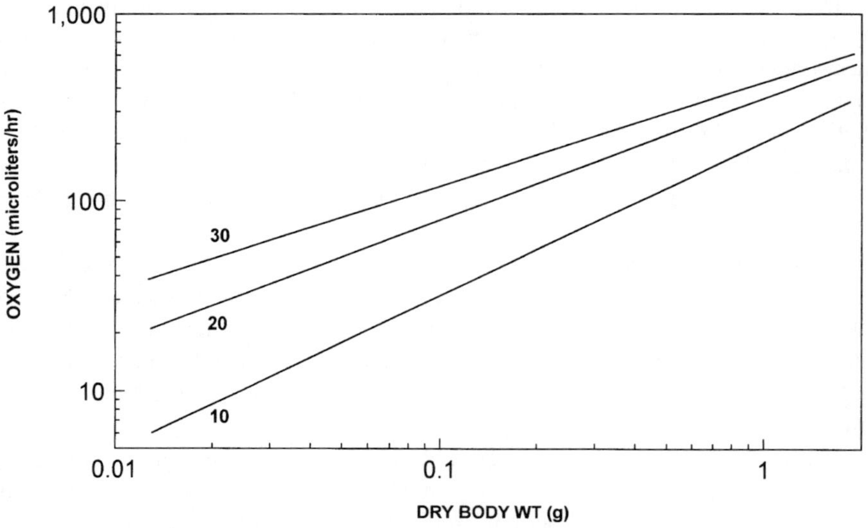

FIGURE 13. The influence of temperature on the respiration rates of intertidal American oysters, *Crassostrea virginica*. (Adapted from Dame 1972.)

acclimated to a temperature regime and be exposed to a variety of temperatures during a tidal or daily cycle. Physiological ecologists have approached this situation by acclimating organisms to a series of temperatures and then observing physiological rates over a range of temperatures for a given acclimation temperature. Initially, only a single environmental factor and a specific physiological rate were observed, but now balanced energy budgets are developed for a target species (Kuenzler 1961 and Dame 1972) and integrated into population level estimates.

Recent work has emphasized the combined effects of complex environmental variables on the integrated physiological response of the organism as a whole. This multidimensional approach, often called "response-surface analysis" (Alderdice 1972), has been applied to a number of bivalves including *Mytilus* (Widdows 1978; Bayne and Scullard 1978) and *Ostrea* (Buxton et al. 1981). In the case of *Ostrea* (Buxton et al. 1981), multiple linear regression analysis was used to obtain polynomial expressions for assimilated ration as a function of acclimation temperature and exposure temperature and respiration of the general form

$$Y = a + bX_1 + b_2X_2 \ldots$$

and these relationships generated a 3-dimensional response surface.

In summary, temperature is a major physical factor that limits the physical extent of bivalves both latitudinally and vertically. This abiotic parameter also influences ecosystems by controlling functional processes, such as survival, growth, feeding, and excretion. As temperature is seasonal in all but equatorial and deep sea habitats, this factor induces a considerable amount of temporal variation on bivalves and their ecosystems.

SALINITY

While temperature is generally recognized as the principal factor at large biogeographic scales, salinity is an important determining factor in the distribution of coastal and estuarine bivalves and is influential on many physiological rates. Salinity may affect the structural and functional properties of animals through changes in (1) total osmotic concentration, (2) relative proportions of solutes, (3) coefficients of adsorption and saturation of dissolved gases, (4) density and viscosity (Kinne 1964). Although salt concentrations or salinities are fairly constant in the oceans at about 35‰, in coastal and estuarine regions salinity varies along a gradient from freshwater in landward areas to full marine salinity. Depending on the geographical location, salinity may vary seasonally due to seasonal river input and freshwater runoff and precipitation. Bivalve molluscs are found at all salinities including freshwater and hypersaline areas typical of tropical zones.

Because estuarine and coastal species are not uniformly distributed across these salinity gradients, numerous schemes have been developed to describe the observed distributions (Bulger et al. 1993). The Remane and Schlieper scheme has been criticized because its database came from the Baltic Sea and probably incorporates a temperature gradient as well as a salinity gradient. However, the latter scheme was documented by Khlebovich (1969) to describe the principle of critical salinity. This principle states that there is a salinity barrier between 5 to 8‰. Above this barrier, poikiloosmotic organisms can survive, but below the barrier, hyperosmotic regulation is required. Thus, this scheme is physiologically based.

The most well-known scheme is the Venice system (Anonymous 1959). The Venice system was constructed from the experiences of the scientists of the day and has no reported objective criteria. The Bulger scheme supports the Venice system, but is based on field data and principal component analysis. The Bulger scheme divides the salinity gradient into the following overlapping zones: 0–4, 2–14, 11–18, 16–27, and 24–marine ‰.

At the system or community level, Wells' (1961) study of the fauna of oyster beds along a salinity gradient clearly demonstrates the influences of salt concentrations. Utilizing a series of nine stations along a 10-mile stretch of the Newport River estuary in North Carolina, Wells found a consistent decline in animal species in the oyster beds from over 60 species at the most saline station to less than 10 species at the most upstream site with salinities less than 10‰. Wells supported his field observations with a laboratory examination of low salinity tolerance by the 20 most common animal species found on these beds, including the bivalves *Crassostrea*, *Mercenaria*, *Geukensia*, *Brachidontes*, and *Chione*. A comparison of salinity death points and distribution along the salinity gradient resulted in a more or less similar ranking for the bivalves and all but two of the other animals. From these results, Wells concluded that a great majority of the species on the oyster bed are limited in their upstream distribution or penetration by salinity.

Salinity also influences many of the major rate processes in bivalves, although laboratory and field studies do not always appear to agree (Bayne et al. 1976). Typically, the respiration rates of euryhaline bivalves are similar when measured at their field-ambient salinities (Remane and Schlieper 1971), but major differences occur when there is an acclimation period. Some of this discrepancy may be due to the necessity for a much longer acclimation period for salinity as opposed to the normal 2-week period for temperature acclimation (Bayne et al. 1976). Changes in environmental salinity may disrupt the steady-state balance between the input and output of cellular water and salts (Hawkins and Bayne 1992). Most bivalves respond immediately to changes in their environment by closing their shells and isolating themselves from the external salinity environment. This isolation helps to reduce the rate of associated changes in cell volume and allows isoosmotic intracellular regulation to commence (Hawkins and Bayne 1992). Thus, following a sudden change in salinity, physiological rates of feeding and respiration are depressed, but gradually the animal regains normal function.

As with temperature, salinity can vary along a spatial gradient, particularly in coastal regions. Temporal variation may also occur due to precipitation and landwater runoff, but these variations are not as obvious as those noted for temperature. Like temperature, salinity limits the distribution of bivalves and can control the basic physiological rates that translate to functional processes in ecosystems.

TEMPERATURE–SALINITY AND OTHER FACTOR COMBINATIONS

From the preceding discussion, it is evident temperature and salinity are significant abiotic factors that have major influences on biological processes. These two factors are often correlated in various ways. Temperature can modify the effects of salinity and enlarge, narrow, or shift the salinity range of an individual, and vice versa (Kinne 1964).

The larvae and adults of bivalves are two distinct morphological and physiological organisms that occupy very different ecological environments (Lough 1974). Because of their coastal and estuarine water column environment, the larvae of most coastal bivalves experience great variations in temperature and salinity during their life stage. Using the response surface approach of Alderdice (1972), Lough (1974) compared the survival and growth of the early and late stage larvae of *C. virginica*, *M. mercenaria*, and *Mulinia lateralis* to the combined influences of temperature and salinity. Although the three species are euryhaline as adults, their larvae have a comparatively narrow salinity range. The observed progressive change in temperature–salinity tolerance with age or time approaches the range tolerated by the adults. Maximum predicted growth occurred at higher temperatures and somewhat higher salinities than those for maximum survival in all three species. Late larval survival and growth appear to be maximized to take advantage of the complex environment the larvae experience on approaching settlement. As bivalves move from one ecosystem to another during their larval stage, it is important that those interested in the whole system be aware of this spatial transition and the role continuously varying abiotic factors may play.

The adult stage of most euryhaline bivalves is usually sessile or semisessile, and the animals must experience many potentially lethal or deleterious abiotic environmental stresses. Not only do temperature and salinity vary, but oxygen and other related chemical parameters may also impact on bivalve survival and physiological rates. For example, Shumway and Koehn (1982) show the effect of declining oxygen tension on the respiration of *C. virginica* adults at different temperatures and salinities, also using the response surface technique. Regulation of respiration in declining oxygen tension decreases with decreasing salinity and increasing temperature. Maximum respiration occurs at moderate salinity and 30°C at all oxygen tensions. In the case of *C. virginica*, this bivalve appears to have a rather "elastic" or "euryplastic" (Alderdice 1972) physiological ecology that allows it to utilize available oxygen over a broad range of temperature–salinity–oxygen combinations without falling into anaerobiosis. In other words, these oysters are well adapted to life in their constantly varying environment.

The interaction of the two factors, temperature and salinity, with various bivalve processes can often describe the potential habitat of a given animal. It is also clear that because of the great differences in the benthic vs. the water column the adult bivalve responses to the physical parameters of temperature and salinity are different from those of the larvae of the same species. These life history differences allow bivalves to exploit and exist in different ecosystems and, thus, increase their chance of survival and influence.

WATER MOVEMENT

The interaction of moving water with bivalves has only recently been recognized as an important phenomenon. Most water flow is created by currents, usually tidal or wind driven, but on open shores or in shallow open bays water flow can also be generated by wave action. Filter feeding bivalves require sufficient water motion to bring in new supplies of suspended food and take away waste. In contrast, deposit feeding bivalves require low water flow to allow the accumulation of deposited seston as a food resource. In this view, the flowing water provides an energy subsidy that allows sessile bivalves to utilize a particular habitat far beyond the ability of the adjacent water column to provide food. If filter feeding bivalves accumulate in large numbers, the structures created by their beds can greatly alter the flow environment and in turn allow bivalves to play a role in the hydrodynamics and development of certain systems.

From an ecological perspective there are two fundamental flow regimes, laminar and turbulent flow. In laminar flow, all the fluid particles move more or less parallel to each other in smooth paths, and this motion is the same at both low and high scale flow. Turbulent flow is very different, with the fluid particles moving in a highly irregular manner even if the fluid as a whole is traveling in a single direction. The transition point between the two flow regimes is often abrupt and is also where biological processes are of major importance (Vogle 1981).

The removal of momentum from a moving fluid by a body, i.e., a bivalve, is known as drag. Because living systems are not uniform objects, drag is difficult to predict for them. As it turns out, the Reynolds number (*Re*) is a valuable descriptor of not only drag, but also other aspects of fluids interacting with solids. The *Re* is a scaling parameter that spans the spatial scales of living systems from molecules to ecosystems. Its role is sometimes comparable to that of the surface-to-volume ratio in physiology (Vogel 1981). Early studies showed that increasing fluid speed, increasing the size of the object in the flow, increasing the density of the fluid, and decreasing the viscosity of the fluid could shift a flow from laminar to turbulent. This combination of parameters derives the dimensionless *Re*.

$$Re = \frac{lU}{\nu}$$

where l is the size of the object, U is the velocity of the fluid, and ν the kinematic viscosity that is the ratio of the dynamic viscosity and density. The drag coefficient, C_D, describes the behavior of drag and is a function of the *Re*.

Because most adult bivalves live on the bottom, it is water velocity near the bottom that is most ecologically important to these organisms. As water flows over the bottom, it is subject to frictional drag with the bottom that dissipates energy (momentum) in the flow and slows down the current. Therefore, at the interface between the current and the bottom the velocity of the water is zero or near zero. The frictional effects of the bottom are transferred within the flow by molecular and turbulent mixing. Thus, there is a velocity gradient from near zero at the bottom to some distance above where the current is not significantly influenced by the bottom. This gradient is called the boundary layer (Figure 14). The boundary layer may be

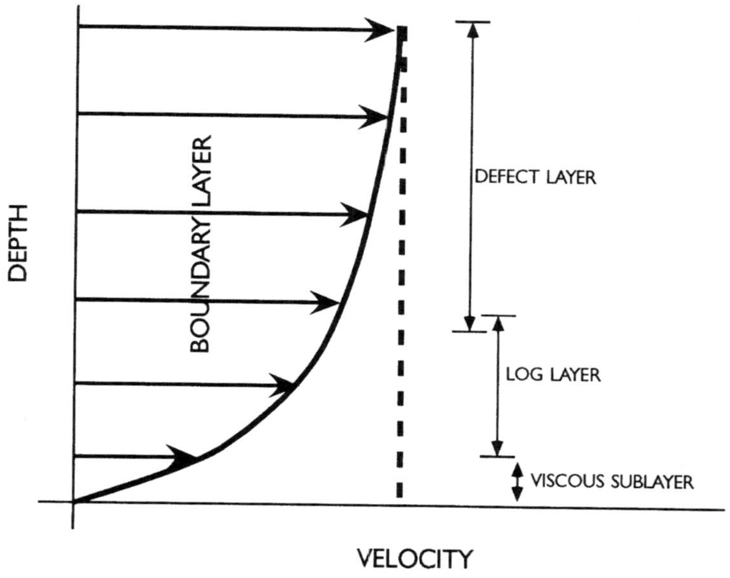

FIGURE 14. A generalized diagram showing the influence of depth and water velocity on the different components of the boundary layer.

either turbulent or laminar. A laminar boundary layer is a region with very little exchange with the surrounding layers of water and often becomes depleted of nutrients and food, while accumulating waste materials. Laminar boundary layers are usually found at low water velocities, and turbulent layers are found at higher water velocities. Bottom roughness can enhance turbulent boundary layer formation and, thus, increase mixing within the current.

In gradually varying flow regimes typical of uneven bottoms and changing water currents, the boundary layer can be divided into three layers (Frechette et al. 1993). Nearest the bottom is the viscous sublayer (Figure 14) where velocity varies linearly with height above the bottom. This layer is continuous and well defined over relatively smooth bottoms, but becomes discontinuous over rougher surfaces like mussel beds and oyster reefs. Under the latter conditions, the flow is said to be rough-turbulent and transport is largely determined by turbulent mixing processes. The next layer is the log layer where velocity varies logarithmically with distance above the bottom. The outermost layer is known as the defect layer and flow behavior here is largely independent of bottom roughness.

Bivalves may respond to flowing water by orienting their shells to reduce drag and, thus, increase stability and increase the availability of food and removal of waste. In sessile bivalves, it is well known that *C. virginica* living in high flow environments have elongated shells while those in low flow conditions have more rounded shells (Kauffman 1971). This observation implies a cause and effect relationship between water flow and structural morphology. The shell surface of many sessile bivalves is often rough with either uniform or nonuniform protuberances. This surface roughness may be an example of the "dimple" or "golf ball" phenomena described by Vogel (1981), where surface roughness induces the stability of an object in a current. From a slightly different perspective, Lawrence (1971) found that individual or isolated *C. virginica* in South Carolina showed little orientation to tidal currents, while densely packed or clustered animals on reefs oriented their planes of commissure parallel to water flow. He also found that fossil beds of *Crassostrea gigantissima* in North Carolina likewise oriented parallel to tidal flow.

Not only do surface dwelling sessile bivalves orient to the flow, but siphonate infaunal bivalves do also. Vincent et al. (1988) noted that *Mya arenaria* usually orients normal to the axis of tidal flow over the clam flats. This orientation suggests that the preferred alignment minimizes the mixing of exhaled and inhaled water, thus optimizing filter feeding and reducing refiltration. From a system perspective, not only do bivalves respond to the boundary layer flow, but they also may also influence the boundary layer by pumping jets and siphon behavior (Monismith et al. 1980). From flume studies, Monismith et al. (1980) showed that a significant portion of excurrent water could be refiltered depending on the boundary layer velocity profile, velocity of the excurrent water and siphon height, orientation, and size. With nonsiphonate epifaunal mussels only the roughness generated by the shells appeared to influence turbulent mixing, and mussel pumping was not a necessary consideration to explain the observed velocity profiles (Frechette et al. 1993).

In surface dwelling byssate and nonbyssate mobile bivalves, the animal tends to respond to a current by orienting the inhalant portion into the current and the exhalant area downstream. In addition, shell shape appears to influence the channeling of water across the bivalve from inhalant to exhalant areas. Evidence from flume experiments (Wildish et al. 1982, 1992) has clearly shown that forced orientation of the exhalant area of scallops into the current resulted in reduced growth and food uptake.

Many scallops have developed jet propulsion or swimming as means of predator avoidance. This phenomenon is an adaptation of shell morphology and internal structure into an integrated behavior for survival. In this behavior, the bivalve not only utilizes the current, but because of its shell shape and jet propulsion generates lift and becomes an object moving through a fluid rather than just responding to the moving fluid (for more on this topic, see Wildish and Kristmanson 1993; Dadswell and Weihs 1990; Morton 1980).

At a higher scale of organization, Grave (1905) hypothesized that oyster reefs build out from the shore towards the central channel. Later work by Keck et al. (1973) linked oyster reef development to tidal creek development. These workers found that reefs are often associated with large creek meanders where higher current velocities on the concave side of the bend result in areas of scour. The sediments of this zone tend to be firm and swept free of soft mud and slime that are unsuitable for spat settlement. The convex side of the meander is an area of lower current velocities and a depositional zone for finer sediments and, thus, generally unsuitable for settlement and oyster survival. Once the reef has been established on the concave portion of the meander, the higher water velocities provide more food and carry away feces and pseudofeces. A recent study by Dame et al. (1987) noted that on a South Carolina oyster reef there was sufficient sediment deposition to cause burial within several tidal cycles. However, in this system maximum ebb tide water velocities over the reef were always adequate to resuspend the accumulated materials.

In addition to structural effects, many functional processes are influenced by water movement. The most well studied of these relationships is the role of current speed on bivalve growth. In this process, it is assumed that the bivalves, particularly dense populations, reduce the concentration of suspended food in the overlaying boundary layer. Initially it was thought that the depletion of food from the boundary layer was a simple function of reduction in current speed (Wildish and Kristmanson 1984), and that this reduction in available food led to reduced growth. Later studies by Frechette et al. (1993) indicated that vertical mixing also plays a role. Depletion of the boundary layer can be alleviated by wave action or currents, but resuspension itself does not appear to positively affect the growth of mussels (Frechette and Grant 1991). Thus, it is appropriate that field observations of mussel beds of different sizes show higher mussel growth along the leading edges and reduced growth along the direction of flow and in the central portions of beds (Newell 1990).

Waves are another form of water flow that influences bivalves. On rocky shores, many bivalves attach themselves to the substrate either through cementation or byssal threads. Although bivalves in these environments benefit from the seston produced by the wave-generated turbulent flows, there is some evidence (Harger 1971) that bivalves facing the direct ocean waves are not as large as those protected by boulders. These observations suggest that energy for growth may be diverted into repairing injuries sustained from more powerful waves.

The role of wave energy in enhancing intertidal productivity has been examined along the northeastern Pacific by Leigh et al. (1987). These authors found that on shores with more wave energy productivity was higher. As intertidal organisms cannot transform wave energy into biochemical energy nor harness wave energy the real question is, how does this energy enhance productivity. Apparently, primary production is enhanced through numerous mechanisms: (1) increased primary production results in more food for animals like bivalves; (2) increased turbulence constantly renews the seston in the area of the bivalves, thus providing more food availability; and (3) wave action can protect residents of the intertidal zone by knocking away their enemies or preventing them from feeding.

Bivalves living in current- and wave-dominated environments are the beneficiaries of increased food availability and waste removal, and these are essentially energy subsidies beyond normal solar inputs to ecosystems. Thus, in these environments it is very common to find large populations of bivalves that probably play major roles in the functioning of their respective ecosystems.

TIDES

At the land–sea interface, tides have the potential to produce major effects on systems dominated by bivalves. The zone between the high and low tide marks is called the intertidal or littoral zone. Organisms in this zone are subject to exposure as the tide drops and submergence as the tide rises. During exposure, inhabitants are faced with increased temperature variability and increased desiccation. During submergence, temperature extremes are minimized and there is no desiccation, but in open areas, wave action can come into play by dislodging individuals or predators.

Although the rise and fall of the tides tracks a smooth curve, zonation in the intertidal region is often attributed to tides. If a series of tidal curves is plotted, a series of distinct breaks is evident, and these breaks are thought to be related to the boundaries between zones. From Figure 15, showing maximum time of continuous submergence for various tide levels along the California coast, certain points or critical tide levels (Doty 1946) reflect sharp increases in exposure to air. These levels have been invoked to explain intertidal zonation. Support for the critical tide level explanation for zonation has been well accepted, most likely, because of the great variations in topography from place to place in the intertidal zone. Other mainly biological factors certainly play a role in intertidal zonation and these will be discussed in Chapter 4.

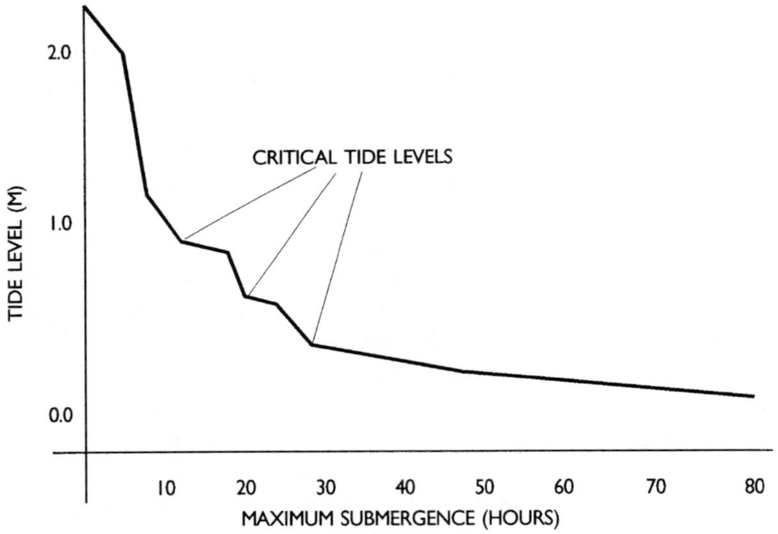

FIGURE 15. A diagram of critical tidal exposure levels along the California coast. (Redrawn after Doty 1946.)

At an ecosystem scale, tides allow the exchange of water between coastal environments. Thus, water molecules within coastal systems are constantly being replaced by other water molecules from either land runoff or the sea. The time it takes for all the molecules of water in a water mass to be replaced is called the turnover time. The time the average water molecule stays within a system is termed the residence time. Turnover times range from hours to months, depending on tidal fluxes, land water inputs, and volume of the system. Short turnover times in the order of hours are typical of low volume systems or systems that go dry at low tide, while long turnover times are found in large volume seas and bays. The longer the turnover time the more time bivalves have to interact with the water volume. However, the biomass of bivalves now becomes an important factor in determining influence. High bivalve densities can offset the tidal dominance of rapid turnover times and lead to control of material turnover rates by these animals (Dame et al. 1992).

In summary, tides can influence the spatial distribution of bivalves in coastal environments. Tides along with coastal geomorphology can determine the turnover time of coastal water volumes, and bivalve biomass can in some cases override the influence of tides in material transport.

SEDIMENTS

Sediment composition is another physical parameter that seems to play a role in the spatial distribution of bivalves. Particle size, as catagorized into sand, silt and clay, is used to describe sediments. Often zones of low energy or little water flow are dominated by finer sediments while areas of high energy are composed of coarser sediments like sand. Since sediments and water flow are correlated, gradients of sediments are often found, particularly in coastal waters and in the intertidal zone. Generally, deposit feeding bivalves are prevalent in finer sediments, and filter feeding bivalves are common to coarser sediments (Rhoads and Young 1970). Although some of this bivalve spatial variation is due to local hydrodynamics and food availability, some is due to the specific species ability to live on the surface of changing sediments or burrow into the sediments. Because sediments are often in particle size gradients, it is not unexpected to find identifiable zones inhabited by different communities of bivalves in intertidal sands and muds (Peterson 1991).

It has long been known that benthic suspension feeders and deposit feeders exhibit reciprocal spatial distributions. In environments where one type of bivalve is abundant and diverse, the other is reduced. Rhoads and Young (1970) found that deposit feeders, especially bivalves, intensively rework surface sediments and produce the following changes: (1) an uncompacted sediment consisting of fecal pellets and reworked material of semiconsoladated mud; (2) a surface of biogenic sand-size particles of low bulk density; and (3) sediments of high water content. These biological modifications of the sediment affect the physical stability of the bottom by increasing interface roughness and lowering critical erosion velocity. This physical instability of the surface sediments is proposed to be stressful to suspension feeding benthos (bivalves) by: (1) clogging filtering structures; (2) resuspending and burying newly settled larvae; and (3) discouraging the settlement of suspension feeding larvae. Unstable sediments also do not provide a suitable substrate for sessile filter feeding bivalves to attach to. Rhoads and Young (1970) further argue that this explanation is limited to areas of high primary production where food is not limiting to bivalve filter feeders. As noted earlier, Wildish and Kristmanson (1993) have observed that, in detailed velocity profile studies of similar habitats, it is the availability of food for benthic suspension feeders in the benthic boundary layer that may be limiting their success. The size and composition of the benthic boundary layer are a result of current velocity, bottom composition, and biotic interactions.

REFERENCES

Alderdice, D.F. 1972. Responses of marine poikilotherms to environmental factors acting in concert, in Kinne, O., Ed., *Marine Ecology*, Wiley, London, 1659–1722.

Anonymous. 1959. Symposium on the classification of brackish waters, Venice, 8–14th April 1958, *Arch. Oceanograf. Limnol.*, 11 (Suppl.)

Bayne, B. L. and Scullard, C. 1978. Rates of oxygen consumption by *Thais (Nucella) lapillus*, *J. Exp. Mar. Biol. Ecol.*, 32, 97–111.

Bayne, B.L., Thompson, R.J., and Widdows, J. 1976. Physiology I, in Bayne, B.L. Ed., *Marine Mussels: Their Ecology and Physiology*, Cambridge University Press, New York, 121–206.

Bourget, E. 1983. Seasonal variations of cold tolerance in intertidal mollusks and their relation to environmental conditions in the St. Lawrence Estuary, *Can. J. Zool.*, 61, 1193–1201.

Bulger, A.J., Hayden, B.P., Monaco, M.E., Nelson, D.M., and McCormick-Ray, M.G. 1993. Biologically-based estuarine salinity zones derived from a multivariate analysis, *Estuaries*, 16, 311–322.

Butler, P.A. 1953. Oyster growth as affected by latitudinal temperature gradients, *Commer. Fish. Rev.*, 15, 7–12.

Buxton, C.D., Newell, R.C., and Field, J.G. 1981. Response- surface analysis of the combined effects of exposure and acclimation temperatures on filtration, oxygen consumption and scope for growth in the oyster *Ostrea edulis*, *Mar. Ecol. Prog. Ser.*, 6, 73–82.

Crisp, D.J., and Ritz, D.A. 1967. Temperature acclimation in barnacles. *J. Exp. Mar. Biol. Ecol.*, 1, 236–256.

Dadswell, M.J. and Weihs, D. 1990. Size related hydrodynamic characteristics of the giant scallop, *Placopecten magellanicus*, *Can. J. Zool.*, 68, 778–785.

Dame, R.F. 1972. The ecological energies of growth, respiration and assimilation in the intertidal American oyster, *Crassostrea virginica*, *Mar. Biol.*, 17, 243–250.

Dame, R.F. 1987. The net flux of inorganic matter by an intertidal oyster reef, *Cont. Shelf Res.* 7, 1421–1424.

Doty, M.S. 1946. Critical tide factors that are correlated with the vertical distribution of marine algae and other organisms along the Pacific coast, *Ecology*, 27, 315–328.

Frechette, M. and Grant, J. 1991. An *in situ* estimation of the effect of wind-driven resuspension on the growth of the mussel, *Mytilus edulis*, *J. Exp. Mar. Biol. Ecol.*, 148, 201–213.

Frechette, M., Lefaivre, D., and Butman, C.A. 1993. Bivalve feeding and the benthic boundary layer, in Dame, R.F., Ed., *Bivalve Filter Feeders in Coastal and Estuarine Ecosystem Processes*, Springer-Verlag, Heidelburg, 325–369.

Galtsoff, P.S. 1964. The American oyster *Crassostrea virginica* (Gmelin), *U.S. Fish Wildlf. Serv. Fish. Bull.*, 64, 1–480.

Grave, C. 1905. Investigations for the promotion of the oyster industry of North Carolina, *Rep. U.S. Fish Comm.*, 1903, 247–341.

Harger, J.R.E. 1971. The effect of wave impact on some aspects of the biology of sea mussels, *Veliger*, 12, 401–414.

Hawkins, A.J.S. and Bayne, B.L. 1992. Physiological interrelations, and the regulation of production, in Gosling, E., Ed., *The Mussel Mytilus: Ecology, Physiology, Genetics and Culture*, Elsevier, Amsterdam, 171–222.

Kauffman, E.G. 1971. Form, function, and evolution, in Moore, R.C., Ed., *Treatise on Invertebrate Paleontology,* Part N, *Mollusca 6, Bivalvia*, Geological Society of America, Lawrence, KS, N129–N210.

Keck, R., Maurer, D., and Watling, L. 1973. Tidal stream development and its effect on the distribution of the American oyster, *Hydrobiology*, 42, 369–379.

Khlebovich, V.V. 1969. Aspects of animal evolution related to critical salinity and internal state, *Mar. Biol.*, 2, 338–345.

Kinne, O. 1964. The effects of temperature and salinity on marine and brackish water animals, II. Salinity and temperature salinity combinations, *Annu. Rev. Oceanogr. Mar. Biol.*, 2, 281–339.

Kuenzler, E.J. 1961. Structure and energy flow of a mussel population in a Georgia salt marsh, *Limnol. Oceanogr.*, 6, 191–204.

Lawrence, D.R. 1971. Shell orientation in recent and fossil oyster communities from the Carolinas, *J. Paleontol.*, 45, 347–349.

Leigh, E.G., Paine, R.T., Quinn, J.F., and Suchanek, T.H. 1987. Wave energy and intertidal productivity, *Proc. Natl. Acad. Sci. U.S.A.*, 84, 1314–1318.

Liebig, J. 1847. *Chemistry in Its Applications to Agriculture and Physiology*, Taylor and Wilson, London, 240 pp.

Lough, R.G. 1974. A reevaluation of the combined effects of temperature and salinity on survival and growth of bivalve larvae using response surface techniques, *Fish. Bull.*, 73, 86–94.

Monismith, S.G., Koseff, J.R., Thompson, J.K., O'Riordan, C.A., and Nepf, H.M. 1980. A study of model bivalve siphonal currents, *Limnol. Oceanogr.*, 35, 680–696.

Morton, B. 1980. Swimming in *Amusium pleuronectes, J. Zool. Lond.*, 190, 375–404.

Newell, C.R. 1990. The effects of mussel (*Mytilus edulis*) position in seeded bottom patches on growth at subtidal lease sites in Maine, *J. Shellfish Res.*, 9, 113–118.

Odum, E.P. 1983. *Basic Ecology*, Saunders, Philadelphia, 613 pp.

Odum, E.P., Finn, J.T., and Franz, E.H. 1979. Perturbation theory and the subsidy-stress gradient, *Bioscience*, 29, 349–352.

Peterson, C.H. 1991. Intertidal zonation of marine invertebrates in sand and mud, *Am. Sci.*, 79, 236–249.

Remane, A. and Schlieper, C. 1971, *Biology of Brackish Water*, Wiley-Interscience, New York, 372 pp.

Rhoads, D.C. and Young, D.K. 1970. The influence of deposit- feeding organisms on sediment stability and community trophic structure, *J. Mar. Res.*, 28, 150–178.

Shelford, V. 1913. *Animal Communities in Temperate America*, University of Chicago Press, Chicago, 368 pp.

Shumway, S.E. and Koehn, R.K. 1982. Oxygen consumption in the American oyster *Crassostrea virginica, Mar. Ecol. Prog. Ser.*, 9, 59–68.

Vernberg, F.J. and Vernberg, W B. 1972. *Environmental Physiology of Marine Animals*, Springer-Verlag, New York, 346 pp.

Vincent, B., Desrosiers, G., and Gratton, Y. 1988. Orientation of the infaunal bivalve *Mya arenaria* L. in relation to local current direction on a tidal flat, *J. Exp. Mar. Biol. Ecol.*, 124, 205–214.

Vogle, S. 1981. *Life in Moving Fluids: The Physical Biology of Flow*, Willard Grant Press, Boston, 352 pp.

Wells, H.W. 1961. The fauna of oyster beds, with special reference to the salinity factor, *Ecol. Monogr.*, 31, 239–266.

Widdows, J. 1978. Combined effects of body size, food concentration and season on the physiology of *Mytilus edulis*, *J. Mar. Biol. Assoc. U.K.*, 58, 109–124.

Wildish, D.J. and Kristmanson, D.D. 1984. Importance to mussels of the benthic boundary layer, *Can. J. Fish. Aquat. Sci.*, 41, 1618–1625.

Wildish, D.J. and Kristmanson, D.D. 1993. Hydrodynamic control of bivalve filter feeders: a conceptual view, in Dame, R.F., Ed., *Bivalve Filter Feeders in Coastal and Estuarine Ecosystem Processes*, Springer-Verlag, Heidelburg, 299–324.

Wildish, D.J., Kristmanson, D.D., Hoar, R.L., DeCoste, A.M., McCormick, S.D., and White, A.W. 1987. Giant scallop feeding and growth responses to flow, *J. Exp. Mar. Biol. Ecol.*, 113, 207–220.

Wildish, D.J., Kristmanson, D.D., and Saulnier, A.M. 1992. Interactive effect of velocity and seston concentration on giant scallop feeding inhibition, *J. Exp. Mar. Biol. Ecol.*, 155, 161–168.

3 Organismic Level Processes

INTRODUCTION

Organismic level processes, e.g., feeding, respiration, excretion, and growth, are often fundamental to examining the structural and functional roles of bivalves at the population and ecosystem levels. Traditionally, these organismic rate processes have been summed, equivalent to scaling up, to yield population and ecosystem estimates. Frequently these processes have been grouped together or integrated into ratios, either indices or budgets, in an effort to provide a deeper understanding of the underlying physiological or ecological processes.

In a more ecological perspective, the energy or material budget (Bayne and Newell 1983; Crisp 1984; Griffiths and Griffiths 1987) has been utilized to further understand the various energy and material-processing relationships of organisms by summing a number of rate processes. The energy budget applies the laws of equilibrium thermodynamics to organisms and populations. Thus, it is assumed that for energy or material:

$$\text{Input} = \text{Output}$$

and that these terms can be expanded to:

$$C = P + R + F + U$$

Where C is the energy or material content of the food consumed; P is that quantity produced as tissues or gametes; R is that lost as metabolic heat or gas; F is the quantity lost as feces; and U is the amount lost from excreted urine and mucus. Production can be further divided:

$$P = P_g + P_r$$

Where P_g is the energy or material concentrated in new tissues and P_r that lost as reproductive material. Assimilation is defined as the amount of energy or material consumed (C) that is utilized by the organism. A can be calculated as:

$$A = C - (F + U) = P + R$$

The concept of "scope for growth" is defined as the difference between the energy of the food the animal consumes and all the other utilizations and losses (Warren and Davis 1967) and is readily derived from energy or material budgets. Such estimates have been applied to economically important bivalves including *Mytilus edulis* (Widdows and Bayne 1971), *Crassostrea virginica* (Dame 1972), and *Ostrea edulis* (Buxton et al. 1981). These and other studies on bivalves indicate that the capacity of bivalves to regulate the various components of the energy budget by acclimation may enhance their ability to maintain a positive scope for growth. Results from such studies also suggest that outside the limits of acclimatory adjustment, energy balance considerations may limit the bivalve's existence to a particular ecological niche (Buxton et al. 1981).

In this chapter, bivalve physiological rate processes will be examined and related to various integrative measures, including budgets. This approach will develop an understanding of how estimates of ecosystem level rates are achieved through scaling up from organismic rate measurements.

FEEDING

From an ecosystem perspective, feeding is the input of energy and materials into the bivalve. A number of quite different feeding mechanisms have evolved in the bivalves. Filter or suspension feeding is one mechanism that allows the bivalve to remove suspended microscopic particulate materials from the water column. Bivalve filter feeders pump or process water through a structural internal filter that usually retains particles irrespective of their food quality. Deposit feeding is typical of infaunal bivalves living in soft sediments. In this type of feeding, particles on the surface or within the substrate, i.e., deposits, rather than suspended sediments are collected by the bivalve and utilized as food. More specialized in their feeding mechanisms are boring bivalves and bivalves with symbiotic algae or bacteria in their tissues.

FILTER (SUSPENSION) FEEDING

As noted earlier, bivalves evolved in shallow coastal seas where there was an abundance of suspended particulate material, usually dominated by phytoplankton. Because most bivalve filter feeders are sessile or semisessile as adults, these animals must pump and filter volumes of water considerably greater than that immediately adjacent to their habitat in order to consume sufficient phytoplankton to survive. The pumping and filtering mechanism used by bivalves has been a point of biological interest for a considerable time and an understanding of the structure and function of this mechanism is crucial to understanding the types of observations made and translated from the organismic level to the ecosystem level.

The Bivalve Filter and Pump

The process of bivalve filter feeding is the result of the combined action of three types of cilia on the gill filaments (Figure 16). The lateral cilia are the main water

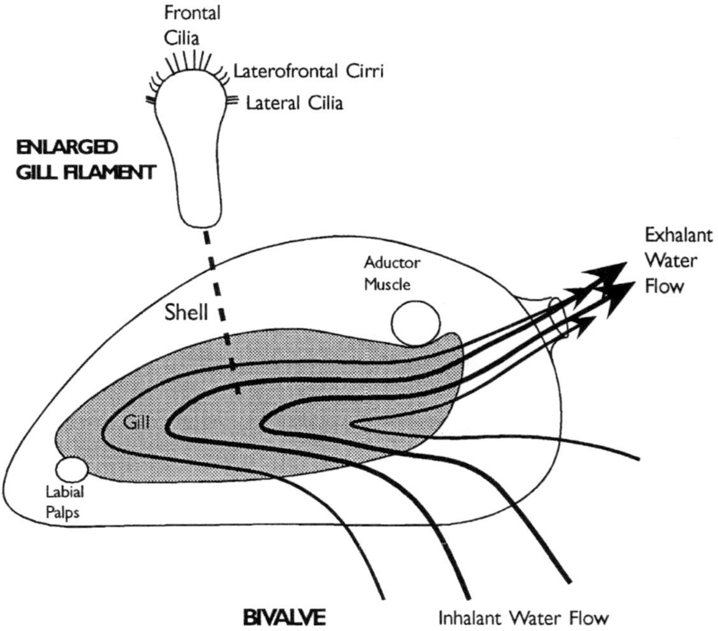

FIGURE 16. Water currents passing through the typical bivalve shell cavity, with a cross-sectional view of a gill filament showing the different types of cilia.

movers and beat with a metachronal wave that assures a minimum distance between the cilia in order to maintain maximum water movement (Sleigh and Aiello 1972). As the water is moved or pumped through the gills, potential food particles are retained by the laterofrontal cirri in a number of ways. The particles may be sieved, intercepted, and impact with the gill through inertia, motile particle motion, or gravitational deposition. Except for sieving, all of these potential ways of particle interception allow for the trapping of particles smaller than the spaces between the cirri of the filter. In addition, particles are carried along in free suspension or caught up in mucus and transported by the frontal cilia in mucus trains or strings towards the labial palps and mouth (Jørgensen 1990; Ward et al. 1993). At the labial palps, particles are sorted either to be transported to the mouth or rejected as pseudofeces.

Although I have presented the bivalve filter as a more mechanical device, Jørgensen (1990) has argued in a controversial fashion that it is a fluid mechanical filter where the complex patterns of laminar currents at the interfilament canals transfer particles across streamlines from through flowing currents to surface currents. Recent *in situ* observations using a fiber optic probe (endoscope) indicate that both mechanical and hydrodynamic filtration mechanisms appear to function in bivalves (Ward et al. 1993) and that these two mechanisms are not mutually exclusive. These authors also directly confirmed that mucus is important in the normal feeding of bivalves. In their study, *Mytilus edulis* used mucus strings on the ventral grooves of the gill filaments, *Placopecten magellanicus* mainly used a hydrodynamic slurry in the dorsal tracts, and *Crassostrea virginica* utilized both mechanisms (Ward et al. 1994).

These workers speculate that the two mechanisms are efficient under different environmental conditions.

The regulation of bivalve filtration rates has also been a contentious issue in recent years. One school of thought (Jørgensen et al. 1986; Jørgensen 1990) maintains that the bivalve filter pump is basically controlled by the physical properties of the filter pump and the water that flows through the system. Jørgensen and his colleagues argue that variations in pumping rate in bivalves in response to temperature can be explained by changes in kinematic viscosity of the flowing water. They base this contention on the correlation between viscosity and filtration rate. Their views do not explain how the pump responds to environmental variation.

The physiological school led by Bayne argues that body size, temperature, physiological condition, and suspended particulate material quantity and quality are important in explaining the variability of bivalve filtration rates. As the filtration rate shows an acute response to changes in temperature, and temperature is most likely a major and fundamental controlling factor on metabolism in poikilotherms (Kinne 1964), Willows (1992) suggests that a general explanation, based on temperature dependence of metabolism, is an underlying cause for filtration rate variability. This view is supported by many studies on the physiological acclimation of filtration rate to temperature (Widdows and Bayne 1971; Widdows 1976). As with temperature controls, the rapid and compensatory decline in filtration rate in response to increased particulate loading is also attributed to physiological processes (Widdows 1976). Thus, a physiological explanation of filtration rates in bivalves seems to be more persuasive at this time.

Filtration and Particle Quality

In natural ecosystems, filter feeding bivalves pump water containing a variety of particles of differing qualitative food values. It is common for suspended inorganic particles of little or no food value to be present in much higher concentrations than phytoplankton and other organic particles. The high concentrations of inorganic particles essentially reduce the food value of the total suspended particulate material. If bivalve filter feeders could qualitatively select particles they could offset the inorganic seston dilution effect. While some research (Winter 1978) asserts that filter feeding bivalves have no qualitative particle selection ability, other long-standing evidence from bivalve stomach contents shows that the particles in the gut are not in the same proportions as those in the adjacent water column (Grave 1916; Morse 1944).

In a series of observations, Kiørboe and Møhlenberg (1981) showed that the efficiency of selection varied from almost nil to 15%, depending on species (Figure 17). The efficiency of selection was clearly related to palp size, but not gill type or turbidity in the respective bivalve's environment. Additional studies by Shumway et al. (1985) concluded that not only are the palps selective, but the gills and stomach may also play a role in particle selection. Other work by Kiørboe et al. (1981) found that sorting did not commence until a minimum concentration or threshold value of suspended inorganic particulate material had been reached. This threshold appears to occur when the concentration of suspended matter exceeds the capacity for

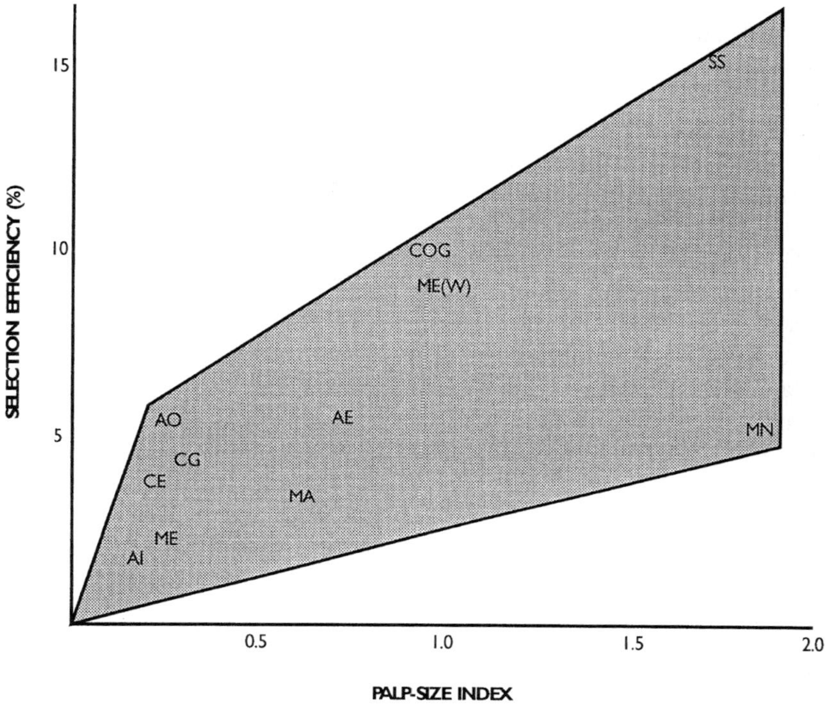

FIGURE 17. The relationship of selection efficiency (as determined from the ratio of relative chl *a* concentration in the inhalent water to the same concentration in the pseudofeces) to the palp-size index (palp area relative to clearance) in bivalves. AE = *Acanthocardia echinata*; AI = *Arctica islandica*; AO = *Aequipecten opercularis*; CE = *Cerastoderma edule*; CG = *Crassostrea gigas*; COG = *Corbula gibba*; MA = *Mya arenaria*; ME = *Mytilus edulis*; ME(W) = *Mytilus edulis* (Wadden Sea); MN = *Musculus niger*; SS = *Spisula subtruncata*. (Adapted from Kiørboe and Møhlenberg 1981.)

ingestion of material retained by the filter. It also appears to be the point at which mucus secretion increases and thus entangles particles to be ejected as pseudofeces (Jørgensen 1990). In *Mytilus edulis,* this threshold value was 1 mg/l of suspended material (Kiørboe et al. 1981). In the same vein, Newell and Langdon (in press) have argued that once sufficient particles are retained on the gill to saturate the gut's capacity to process them, one of two things must happen: (1) the clearance rate must decline or (2) excess particles must be rejected as pseudofeces. This makes sorting a very advantageous process as the gut can be filled with very nutritious particles.

Particle size and quality also play a role in filtration and feeding. Particles that are too large are not utilized as food (rejected as pseudofeces) and particles that are too small may pass through the filter. In addition, particles may have different nutritional values, depending on size. For example, bacterial cells are much higher in numerical concentration than phytoplankton cells in most coastal systems, but bacteria are not as good a source of carbon as phytoplankton (Newell and Field 1983).

Generally, algae are the main source of nutrition for bivalve filter feeders, but in coastal habitats, bacteria, detritus, and nanozooplankton may also make contributions (Langdon and Newell 1990). Several stable isotope studies have shown that both oysters, *Crassostrea virginica*, and marsh mussels, *Geukensia demissa*, utilize *Spartina* (marsh grass) detritus and phytoplankton (Peterson et al. 1985; Kreeger et al. 1988; Langdon and Newell 1990). Interestingly, the further marsh mussels are away from the ocean, the major source of phytoplankton, the higher the potential percentage of detrital carbon in their bodies. These studies suggest that the consumption and assimilation of particles is related to particle abundance and that detrital carbon is assimilated into bivalve tissue. Stable isotope studies are useful because they provide an integrated measure of the contribution of foods of different isotopic signatures to consumers, but they provide little information on the physiology and behavior of the animal regarding feeding.

Other studies, using radioactively labeled *Spartina alterniflora* carbon that had been ground into small particles and fed to bivalves, indicate only a very low absorption (2 to 3%) of ^{14}C in *Crassostrea* (Crosby et al. 1989) and *Geukensia* (Kreeger et al. 1988). Based on these estimates, this form of detrital carbon provides only 0.7% and 8.6%, respectively, of the metabolic requirements for the two bivalve species. When the cellulose assimilation efficiencies of intertidal and subtidal marsh mussels are compared, the intertidal bivalves have a higher efficiency. This higher assimilation efficiency has been attributed to prolonged gut residence time for the intertidal form (Kreeger et al. 1988) and supports the suggestion of Bayne et al. (1988) that intertidal suspension feeding bivalves may physiologically compensate for reduced feeding periods by increasing gut residence time. Not all detrital carbon is as refractory as *Spartina* lignocellulose; thus one might expect a greater percentage of celulosic material being utilized by bivalves in systems with sea grasses or macroalgae. For example, Stuart et al. (1982) investigated the absorption of kelp detritus by the ribbed mussel *Aulacomya ater.* They found that kelp detritus was absorbed with an efficiency of about 50% while kelp bacteria were absorbed with an efficiency of about 70%. These findings suggest that kelp debris may represent an important source of carbon for filter feeders in and adjacent to kelp beds.

Bacteria may also provide, both directly and indirectly, nutrition to filter feeding bivalves. Bacteria can be utilized as food by adult bivalves. Zobell and Landon (1937) and Zobell and Feltham (1938) fed *Mytilus californianus* an exclusive diet of bacteria for 9 months. After that time, the bacteria-fed mussels showed a weight gain of 12.4%, while the unfed controls lost 16.3% of their original body weight. Prieur (1981) examined the digestive tract of *M. edulis* and found partially digested bacterial cells throughout the gut, and undamaged cells were found in the hind gut 3 h after feeding. Prieur (1981) also showed that the bacteria could divide several times during their passage through the gut, so that the yield of bacteria in the feces could be considerably in excess of the biomass of the bacteria ingested as food. The uptake of several species of bacteria labeled with radioactive tritiated thymidine by *M. edulis* was investigated by Birbeck and McHenery (1982). They concluded that *Mytilus* could select for bacteria it was capable of lysing and discriminate between bacterial polymers in order to reject DNA. Seiderer et al. (1984) examined the crystalline style of *Choromytilus meridionalis* in a kelp bed environment and found that this digestive structure contained bacteriolytic enzymes capable of lysing a majority of the free-living suspended bacteria in the surrounding water column. They also observed that style enzymatic activity was highest when bacteria were the predominant exploitable food resource. As the bacteria in this environment have a high carbon:nitrogen ratio (3.7:1), bacteria are a potentially significant source of nitrogen for these mussels. Studies from Chesapeake Bay indicate that *Crassostrea virginica* and *Geukensia demissa* were able to filter free unattached bacteria with an efficiency of 5% and 15.8%, respectively (Kreeger et al. 1988). The latter value is similar to that estimated by Wright et al. (1982) for marsh mussels in a New England estuary. Using Crosby's (1987) 57% assimilation efficiency for bacterial carbon, Langdon and Newell (1990) estimated that unattached, free bacteria could contribute 3.4% and 25.8% to the metabolic carbon requirements of oysters and marsh mussels, respectively. Thus, bacteria in free suspension appear to make a much larger contribution to the carbon budget of mussels and to the nitrogen budgets of bivalves in general.

Suspended organic detritus often has bacteria attached or associated with it and these bacteria not only are involved in the decomposition of the detrital carbon, but are also assimilating dissolved organic carbon released from the detritus. The bivalve filter can remove the larger organic detritus particles with the attached bacterial cells. Crosby (1987) and Crosby and Newell (1990) have shown in laboratory experiments that *C. virginica* can assimilate radioactively tagged cellulolytic bacteria and cellulose detrital complex with an efficiency of 10.3%. This efficiency is almost 4 times the efficiency with which oysters assimilated ^{14}C-labeled cellulose alone. Langdon and Newell (1990) have extended these estimates to summer field conditions in Chesapeake Bay and estimated that attached bacteria could contribute 2.1% of the metabolic carbon requirements of subtidal oysters and 5.2% of the requirements of intertidal marsh mussels. They caution that the actual contribution of attached bacteria is probably even smaller as the total carbon requirement of bivalves is generally about 33% greater than the metabolic requirement (Bayne and Newell 1983).

Langdon and Newell (1990) have argued that nanozooplankton may be a more important means of transferring detrital carbon through the decomposer community to filter feeders. These zooplankters feed primarily on bacteria and because they are 2 to 20 μm in diameter they can be more efficiently retained on the bivalve gill than the much smaller bacteria. Thus, nanozooplankton may be an important link in the transfer of nutrients from bacteria to bivalves (Sherr et al. 1990). Langdon and Newell (1990) have estimated that nanozooplankton could contribute 15% and 37% to the summer metabolic carbon requirements of oyster and mussels respectively at their study site in Chesapeake Bay.

Dissolved Organic Matter (DOM)

Dissolved organic matter (DOM) has long been hypothesized to play an important role in the nutrition of aquatic animals (Putter 1909). Because most DOM can be extremely complex and difficult to characterize, studies of bivalve uptake of DOM have focused on relatively simple dissolved free amino acids (DFAA). Bivalve filter feeders pump large quantities of water and thus are exposed to sufficient quantities of DFAA to make them a potentially significant source of nutrition. Marine bivalve filter feeders, including *Mytilus, Geukensia, Rangia, Cerastoderma, Mya, Ostrea,* and *Bankia,* have been observed to have the ability to transport DFAA (Wright 1982). Uptake of DFAA appears to take place on the surface tissues of the gills, and the system is capable of providing a net input of amino acid from submicromolar concentrations in the surrounding water column. Wright (1982) and Stephens (1982) have calculated that this pathway can supply reduced carbon at rates as high as 60% of the oxidative requirements of the bivalves. The utilization of exogenous amino acids as a nutritional source was challenged by Wright (1985), who contended these DFAAs were only taken up for use in osmoregulation. Later studies by Rice and Stephens (1987) with radiochemical analysis found that as many as 10 different amino acids are taken up from the water column and that these DFAAs are distributed rapidly into the internal tissues of *Crassostrea gigas* and *Mytilus edulis*. All of the preceding findings are from laboratory observations and caution should be taken in applying them to the field. Newell (personal communication) has pointed out that natural marine or estuarine bacteria, not the cultured bacteria used above, most likely can take up DFAA more efficiently than bivalves. Clearly field studies need to be undertaken on the uptake of DFAAs by bivalves *in situ*.

Filtration and Particle Size

Early studies of particle retention (percent removal) by bivalves showed that particles below 1 to 2 μm were usually passed through the animal, while particles larger than 3 μm were normally retained at a higher percentage (see Jørgensen 1966 and 1990 for reviews). Jørgensen (1966) further speculated that the sharp size limit between retainable and nonretainable particles is an indication that it is particle size and therefore the pore size of the filter that is a major factor regulating retention. Atkins (1938) measured the distance between the adjacent cilia on the laterofrontal cirri on a gill filament in *Crassostrea virginica* to vary between 1.5 and 3.7 μm. This range coincides with the inflection point in the observations by Haven and Morales-Alamo (1970), also on *C. virginica*. In a comparative study of filtration by *Mytilus edulis*, *Geukensia demissa*, and *Mya arenaria*, Wright et al. (1982) found that only *Geukensia* was capable of efficiently filtering bacteria. Measurements of the distance between laterofrontal cilia in *Mytilus* and *Geukensia* revealed closer spacing and more overlap of cilia in *Geukensia*. Thus, it appears that the distance between the laterofrontal cilia is one of the determinants of the smallest size particles that can be completely retained on the bivalve filter and allow bivalves like *Geukensia* to utilize smaller cells like bacteria.

Later studies by Jørgensen (synthesized in 1990) and others have shown that although particle size and filter sieve size are important, other properties including water flow within the bivalve can regulate particle capture. Ancient bivalves probably used cilia to generate a respiratory water current to clean the gills of particles retained in this process (Yonge 1947). In such systems, mucus was used to bind particles and transport them along rejection pathways. The currents produced by the gill cilia are laminar and, in conjunction with the gill wall, act to prevent particles from contacting the wall and escaping the current (Jørgensen 1990). In addition to the particle sieve size of the gill, mucus binding and the hydrodynamics of laminar water flow across the gill play a role in the capture and retention of particles. Consequently, the structural and functional features of bivalves as well as the dynamics of water flow act in concert to regulate filtration.

Filtration Models

Filtration or clearance rate is defined as the volume of water completely cleared of particles per unit of time (Bayne et al. 1976). Filtration rate should not be confused with pumping or ventilation rate, which is the volume of water flowing through the gills per unit of time. The two rates are the same if all particles are removed by the bivalve filter feeder's gills, i.e., 100% retention efficiency (Bayne et al. 1976). In reality, the filtration rate is usually less than the pumping rate, particularly as particle size declines below 2 μm.

In practice, feeding rates have been estimated both directly and indirectly. Direct methods usually involve the interception and measurement of the inhalant and exhaling currents of the bivalve (Coughlan 1969). In this approach, the concentration of the particles is determined and multiplied by the water flow to estimate material fluxes. Therefore, the pumping rate (PR) is:

$$PR = (F_i \times C_i) - (F_o \times C_o)$$

where F_i and F_o are the input and output flows and C_i and C_o are the input and output concentrations. This direct approach to estimating pumping rate is difficult to use and is almost always carried out in the laboratory.

Indirect methods emphasize the rate of removal of particles by a bivalve from a known volume of suspension (Coughlan 1969). The calculated rate is a function of the feeding current and retentivity and is termed the filtration rate. Initially, researchers used a static chamber or container whose volume did not change (Coughlan 1969). At least four assumptions are fundamental to calculating the filtration rates in static systems:

1. The reduction in concentration of particles is due to filtration by the animal and gravitational settling.
2. The animal's pumping rate is constant through time.
3. Particle retention is 100% efficient; alternatively, a known constant percentage is retained.
4. The test suspension is homogenous through time.

If all the assumptions hold, the calculated rate is the pumping rate (Coughlan 1969).

Suspension feeding bivalves continuously remove particles from the known water volume and dilute the suspension. If the filtration rate remains constant, the concentration in the chamber will decline exponentially. Clearance can be described by a general relationship:

$$C_t = C_0 \cdot \exp^{\left(\frac{mn}{M}+a\right)\cdot t}$$

where C is the concentration at time t; M is the volume of the suspension; n is the number of animals; m is the filtration rate of a single animal; and a is the rate particles settle out of suspension. Solving for m this relationship becomes:

$$m = \frac{M}{n}\left[\frac{\left(\log_e C_0 - \log_e C_t\right) - a}{t}\right]$$

Coughlan (1969) has shown that the above relationship is equivalent to no less than six previously published clearance equations utilizing different notation. Coughlan also provides a simple graphical method to determine filtration rate in static systems when the sampling interval and the ratio of initial and observed concentrations are known.

The static chamber exponential filtration rate model has been mathematically evaluated by Williams (1982). Williams found that the exponential model is a sufficient description of pumping rate only when filtration efficiencies are 100%. As noted earlier (Haven and Morales-Alamo 1970; Vahl 1972; Møhlenberg and Riisgard 1978; Palmer and Williams 1980), filtration efficiency in bivalves declines for smaller particles, particularly those below 3 to 4 μm. These observations suggest that forms of particle capture other than sieving may be involved in bivalve filter feeding Williams (1982). Rubenstein and Koehl (1977) theorize that particles too small to be retained at 100% efficiency are captured by direct interception, inertial impaction, or random collision with cilia on the bivalve filter. Such filtration of small particles, termed aerosol filtration, theoretically should result in a complex or double exponential decline in cell concentration that does not easily translate into filtering or pumping rates (Williams 1982). It is also very difficult to measure.

The static chamber method of determining filtration rate has evolved in recent years to include automatic maintenance of particle concentration. This approach allows the estimation of filtration rate from the number of particles added to the volume (Winter 1969, 1970, 1973). This closed chamber system can produce variable results because as the experiments progress oxygen concentrations decline and metabolic waste increases. As with the static chamber estimates of filtration rate, Williams (1982) has shown that the change in the proportion of 100% efficiently filtered cells to those less efficiently filtered produces an artifact in the filtration rate estimate — an artifact not related to the physiology of the bivalve, but to the physical size of the cells the bivalve is feeding on. Today, clearance rate is defined as the volume of water cleared of particles large enough to be retained with maximum efficiency (usually 2 to 3 μm). In such cases, the Coulter counter is set to count only particles larger than this size range. In this case, small particles do not interfere with the measurement of clearance rate.

With the advent of automatic particle counters that allow much more rapid and accurate determinations of particle concentrations, flow-through filter feeding systems have become feasible. These systems allow longer time frame measurements, steady-state concentration of particles, and potential experimental manipulations; maintain oxygen concentrations; avoid metabolic waste build up; and can be adapted to field conditions. In these systems, filtration rate is computed as:

$$FR = \frac{[IN]-[OUT]}{[IN]} \times FLOW$$

where FR is the filtration rate, IN is the input particle concentration, OUT is the output particle concentration, and FLOW is the rate of water flow through the system. The animals in the chamber are actually filtering from a concentration that may be different from IN, particularly at low flow rates (Walne 1972; Bayne et al. 1976). Thus, sufficiently high flow rates should be chosen to avoid this problem, or the adjustment recommended by Hildreth and Crisp (1976) should be taken into account. This adjustment uses the concentration of particles in the immediate area of the bivalve in the denominator of the relationship instead of the input concentration.

Although the preceding formulations provide estimates of filtration rate, the rates derived are difficult to scale up to populations or communities of organisms. This difficulty arises because bivalves filter at different rates depending on biotic variables, i.e., size, and physical variables, i.e., temperature. Size is most often modeled as an allometric relationship of the general power function:

$$FR = aW^b$$

where W is the dry weight of the soft body and a and b are the intercept and exponent, respectively. The values for b usually vary between 0.67 and 1.00, indicating a proportionality between surface area and body weight (Winter 1978). As an alternative to weight, shell length is often substituted, but such linear measures are usually considered less accurate than weight measures (Dame 1972; Winter 1978). It should also be noted that models built from animals feeding on a single type of particle, monoculture of phytoplankton or inert suspended material, overestimate filtration when compared to models developed from bivalves feeding on a variety of suspended particles (Doering and Oviatt 1986). These authors offer the only example of combining temperature and size to estimate filtration rate. Their model is:

$$FR = [(L^{b1})(T^{b2})]/a$$

where L is length and T is temperature in °C. By adding temperature to their model, Doering and Oviatt (1986) increased the amount of variation explained by about 25%.

In general, filtration rate decreases as suspended particle concentrations increase (see Winter 1978 for review). Such observations imply that bivalves regulate the amount of water from which particles are removed in relation to food concentration. Winter (1978) provides a schematic representation of the relationships between food concentration and filtration rate (Figure 18). In this scheme, as food concentration

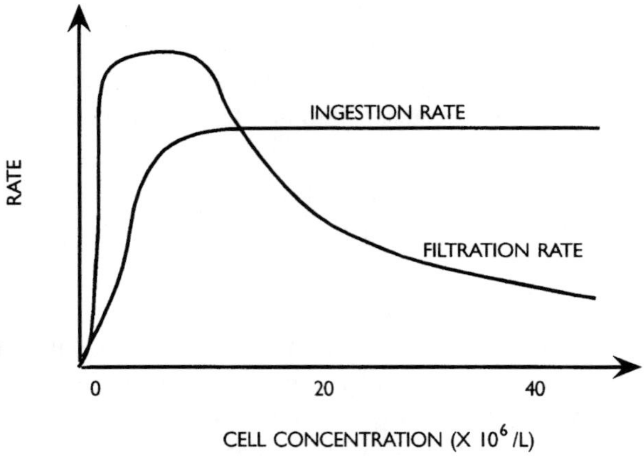

FIGURE 18. A generalized representation of filtration rate and ingestion rate as a function of suspended cell concentration. (Redrawn after Winter 1978.)

rises from some threshold value, filtration rate increases rapidly and stays constant until a maximum of food is ingested. As soon as this maximum ingestion rate is reached, the filtration rate decreases continuously, while food ingestion remains constant. This pattern remains unchanged until the adequate food concentration is reached, after which pseudofeces production begins. At still higher food concentrations, however, filtration and ingestion rates are drastically reduced. These relationships will vary according to particle size, with smaller particles needing higher concentrations to be equivalent to larger particles.

The ingestion rate of particles by bivalves is also a function of body size, temperature, and food concentration. As long as pseudofeces are not produced, the amount of food ingested is equal to the amount of food filtered (Winter 1978). This level of particle concentration is important in aquaculture and is known as the "pseudofeces-free cell density." This concentration appears to be very close or equal to the optimum food concentration, because of the low energy demand for filtration and the high filtration efficiency of algal cells (Winter 1978).

Recently, Willows (1992) developed a functional model of the feeding behavior and physiology of filter feeding bivalves. This model assumes that the net rate of energy gain from the available food sources is maximized. The model is different from previous empirical and physical models in that it optimizes for feeding mainly through physiological adaptation to variations in food quality.

DEPOSIT FEEDING

Deposit feeding bivalves generally inhabit muddy low energy environments, while suspension feeders are more commonly associated with coarser sediments in more energetic environments dominated by tidal and wave-induced currents. Compared with filter feeding bivalves that pump large quantities of water and filter this water for small quantities of suspended particulate food, deposit feeding bivalves not only pump water, but also remove large quantities of deposited sediments from the surrounding benthic environment in order to gain a relatively low percentage of quality organic material. Deposit feeding bivalves meet most of their nutritional needs from the organic portion of the sediments they ingest. Some bivalve deposit feeders can switch back and forth from suspension feeding to deposit feeding or feed in both modes simultaneously, depending on the conditions in their environment. Thus, the classification of a bivalve's status as a deposit feeder may be relative.

Bivalve deposit feeders use the same pumping mechanism as is used by bivalve filter feeders to move water through the animal. Unlike the surface-dwelling filter feeders, many, but not all, deposit feeding bivalves have lengthy incurrent and excurrent siphon tubes that insure the continuity of flow from the benthic–pelagic interface to the animal buried in relatively unconsolidated muddy sediments. For example, in the clam *Scrobicularia*, the inhalant siphon extends and moves across the sediment surface during low tide sucking up surface particles, but retracts during high tide to avoid predation by fish and crabs (Hughes 1969). In the more primitive bivalves, e.g., *Nucula, Solemya*, and *Yoldia*, there is little siphonal development and the animal moves through the muddy sediments ingesting sediments through the incurrent opening. For example, the file clam, *Yoldia*, uses its siphons for water flux, but not feeding. Instead, it uses two grooved tentacle-like appendages to feed. The cilia on these grooves move particles to the clam's mouth (Levinton 1982).

Because the food quality or organic content of sediments is very low, deposit feeding bivalves have developed two approaches to feeding: bulk feeding and sorting. In bulk feeding, large volumes of sediments are processed through the digestive tract in order to gain a small amount of nutrition. Bivalve deposit feeders typically sort particles before they are ingested into the mouth and reject the majority of the particles as pseudofeces (Lopez and Levinton 1987). While there are probably no completely nonselective deposit feeding bivalves, the process of sorting quality food particles from the great mass of poor quality particles results in a lower feeding rate and, in turn, a longer residence time for the food in the gut. As noted earlier for filter feeders, a longer gut residence time enables the digestion of the more refractory organic compounds common to the benthic environment. Thus, deposit feeding bivalves can process up to 20 times their body weight in sediments per hour with as much as 90% of these sediments egested as pseudofeces (Lopez and Levinton 1987).

As deposit feeders feed on the sediments, a number of fundamental question arise. What are the sources of deposit feeding bivalve nutrition? Which components of the sediments are more important as food sources? In this mixture of particles, how does the bivalve separate quality food from poor food?

Microbial Stripping

Studies by Newell (1965) first touted the importance of microbes as food for deposit feeding bivalves. He concluded that *Macoma balthica* ingested both organic material and microbes, but only digested the microbes. This process is referred to as "microbial stripping." Newell (1965) further suggested that the egested organic particles were recolonized by microbes after egestion, and these particles could again be ingested and stripped of microbes, thus making the deposit feeder a major component of a decomposition feedback loop in the sediments.

Cellulase Activity and Detritus Feeding

Although the microbial stripping idea focused on the microorganisms as food, it left unclear the role of organic matter as a food source. One approach to examining the role of organic matter as food for bivalve deposit feeders was to determine if the animals produced cellulase enzymes in their gut. In a survey of bivalve species, Crosby and Reid (1971) found that in both suspension feeding and deposit feeding bivalves, cellulytic activity generally corresponds to the level of cellulose in the food. In the typically deposit feeding genera investigated, *Yoldia* and *Macoma*, cellulytic activity was comparatively low. Crosby and Reid (1971) suggested suspension feeding bivalves have higher cellulase activities than deposit feeders because suspended food includes dinoflagellates that have a cellulose cell wall. Thus, if cellulose is not a general food for deposit feeding bivalves, microbes must be.

Particle Selection

In deposit feeding bivalves, the initial selection of particles occurs with the location and movement of the incurrent siphon or with palp tentacles. *Yoldia*, for example, can use its tentacles to reject up to 95% of the sediment collected (Bender and Davis 1984). As with suspension feeding bivalves, deposit feeding bivalves can also use the size of their filter to select particles. Both tentacles and filters appear to select for finer particles rather than larger particles and this process favors increased particle surface area and microbes. In addition to these selective mechanisms, many tellinid bivalves, e.g., *Macoma* and *Scrobicularia*, have complex stomachs capable of sorting and retaining smaller and less dense particles as opposed to larger particles (Hylleberg and Gallucci 1975). Some of these bivalves may also be adapted for stripping microbes from the ingested particles (Lopez and Levinton 1987). Finally, in some deposit feeding bivalves, the bacteria in the stomach are able to alter the composition of the organic matter in the stomach. This observation led Hylleberg and Gallucci (1975) to suggest that in *Macoma nasuta* the stomach may be compared to a chemostat, where dilution rate and multiplication rate of bacteria are balanced, yielding a net nutrition gain to the bivalve.

Proportion of Foods

Like strictly suspension feeding bivalves, deposit feeding bivalves remove phytoplankton, microzooplankton, organic and inorganic particles, and microbes, including bacteria, fungi, and microalgae from the sediments and surrounding waters. They also probably absorb dissolved organic materials in much the same manner as their filter feeding relatives.

The determination of the relative proportions of the various food sources for deposit feeding bivalves is a major problem (Lopez and Levinton 1987). Some success has been found in combining the ^{14}C-formaldehyde technique of Lopez and Crenshaw (1982) with the ^{14}C:^{31}Cr method of Callow and Fletcher (1972) to estimate the absorption of organic materials. As each of the major food resources for deposit feeding bivalves has a different stable isotope ratio, the use of multiple stable isotope analysis to determine sources of materials and trophic status may also have utility in these endeavors (Peterson et al. 1985).

Optimal Foraging For Food In The Sediments

The concept of optimal foraging is probably so important to the survival of bivalves that extant forms may be living close to their preferred optima (Hughes 1980). As bivalve deposit feeders are not very motile, their foraging strategies are mainly a function of ingestive and digestive processes (Taghon and Jumars 1984). Thus, foraging models (Figure 19) for these organisms stress ingestion and time dependent absorptions of food (Lopez and Levinton 1987). Because of the fine particle nature of the sediments, particle selection is better modeled stochastically (Jumars et al. 1982) and the range of food value is broad, varying from bacteria and digestible plant debris to undigestible materials. As microbial organisms and organic matter are inversely correlated with sediment grain sizes (Yamamoto and Lopez 1985), it is expected that a model of optimal ingestion will predict selection for fine particles (Taghon and Jumars 1984). Microbial feeding with the concurrent production of feces (fecal pellets) may actually stimulate microbial growth through maintenance of logarithmic growth of the microbe populations via grazing and by enriching the sediments with nutrients as they pass through the gut. As a bivalve increases its feeding rate, more food reaches the gut, but gut passage time and digestion rate decrease. This balance of digestion and food reaching the gut sets up an optimal feeding rate that should increase with increasing food quality. Alternatively, Calow (1975) has argued that as food quality decreases, gut residence time should increase to allow increased digestion. Although these strategies seem at odds with each other, if the gut length is longer for bivalves feeding on poorer quality foods, then the ingestion rate need not change. This latter strategy is observed in deep sea bivalves (Allen and Sanders 1966).

SHIPWORMS

Several families of bivalves are specialized for burrowing or tunneling into hard substrates. Species in the genera *Bankia, Lyrodus, Martesia*, and *Teredo* are specialists at boring into wood and are commonly called shipworms. Because these bivalves eventually destroy their habitat, they have developed a number of adaptations that enhance their success in their transient environment. These include: (1) a shell that is an effective rasping tool; (2) a fast growth rate to take advantage of the transient nature of their habitat; (3) early maturation of gonads; (4) high fecundity; (5) incubation of larvae; (6) protandry; and (7) enzymes for cellulose breakdown (Morton 1978).

While all shipworms maintain their ability to filter feed, it is the development of what appears to be a step-wise process of utilizing cellulose derived from wood that has allowed these bivalves to exploit their special habitat in the marine environment. First the shell rasps a tunnel into a piece of wood; then the wood particles

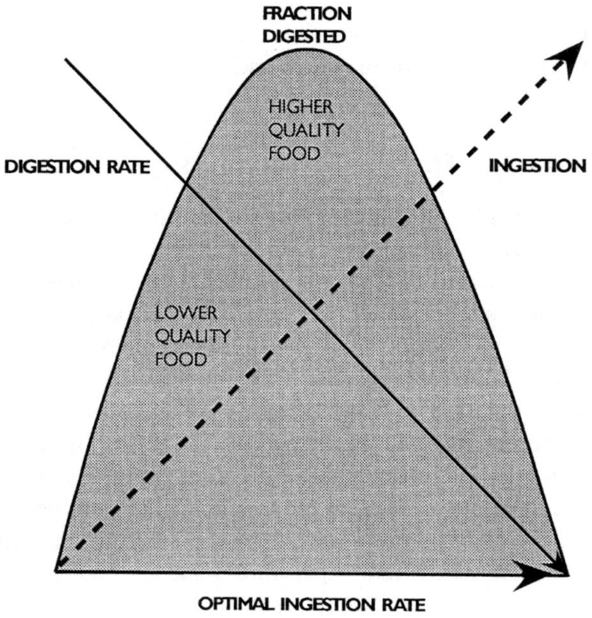

FIGURE 19. The hypothetical relationships between optimal ingestion rate, increasing ingestion, decreasing digestion, and food quality in deposit feeding bivalves. (Redrawn after Lopez and Levinton 1987.)

are taken into the mantle cavity and through a selective sorting mechanism the small fragments are swallowed; in the esophagus, the organ of Deshayes with its symbiotic bacteria secretes cellulytic enzymes; in the Teredinidae, the stomach further sorts the wood particles and additional cellulytic enzymes are produced; finally, the products of lignocellulose decomposition are absorbed by the digestive diverticula and digestion continues intracellularly. Thus, as is typical of other bivalves, digestion is a two-step process with an extracellular digestive sequence and an intracellular sequence (Morton 1978).

Although a similar mucous-producing organ is found in other molluscs, the organ of Deshayes is unique to the shipworm bivalves. It not only produces digestive enzymes that aid in the processing of cellulose, but it may be the only case to date in the animal kingdom of nitrogen fixing bacteria in pure culture in a particular organ (Waterbury et al. 1983).

In summary, shipworm bivalves can play a major part in the decomposition of woody materials in the sea. During the process of adaptation to this role, they may have developed a truly unique symbiotic relationship with nitrogen fixing bacteria.

SYMBIOTIC NUTRITION

There are two fundamentally different types of autotrophic symbioses between bivalves and microbes. In one type, the relationship is between a small number of tropical bivalve species found in coral reef environments and the dinoflagellates,

Symbiodinium (=*Gymnodinium*) *microadriaticum*, also known as zooxanthellae. These autotrophic symbioses are driven by solar energy mediated by photosynthesis (Yonge 1936).

The second type of symbiosis is between a variety of bivalves and bacteria living in the bivalves' tissues. In these chemoautotrophic relationships, the bacteria are chemotrophic, obtaining energy from chemical sources; lithotrophic, because their energy source is inorganic; and autotrophic, because their source of cellular carbon is inorganic (Fisher 1990). As some of these bacteria utilize methane as an energy source and because their cellular carbon source can be organic, these forms are called methanotrophic. While the first bivalves exhibiting chemotrophic symbioses were found around deep sea hydrothermal vents, a great variety of bivalves from mainly anaerobic sulfide-rich environments have now been observed to exhibit chemoautotrophic and methanotrophic symbioses.

Photoautotrophic Symbioses

Members of the tropical giant clam family, Tridacnidae (*Tridacna gigas, T. maxima, T. derasa, T. squamosa, T. crocea, Hippopus hippopus* and *H. porcellanus*) as well as the heart cockle, *Corculum cardissa*, have symbiotic relationships with the zooxanthellae. Almost all of these bivalves are associated with coral reefs where the same species of zooxanthellae is symbiotic with many massive reef building (hermatypic) corals. As with the massive deposition of coral matrices composed of calcium carbonate, the bivalves containing zooxanthellae have a tendency to massive calcium carbonate shells. One species, *T. gigas*, is known as the giant clam and has been reported to exceed 1.2 m in length and 250 kg in weight (Yonge 1936), making it the largest living bivalve. Only certain extinct species of rudist exceed the tridacnids in size, and they also may have had symbiotic zooxanthellae in their tropical/coral reef habitat (Kauffman and Sohl 1974).

Early observations on these large coral reef bivalves quickly focused on the absence of sufficient planktonic food to support their metabolism (Yonge 1936). Furthermore, these bivalves appeared to absorb dissolved inorganic compounds. Various microscopic studies showed that unicellular zooxanthellae were living in the hemal sinuses of the hypertrophied siphon of these bivalves or what is commonly referred to as the "mantle tissue" in tridacnids (Yonge 1936) and to a lesser extent in the other organs (Goreau et al. 1973).

Unlike corals where zooxanthellae are within the cells of the corals, these algae are found intercellularly in the bivalves, which means the larval bivalves must be infected by the zooxanthellae for the symbioses to take place. This uptake of zooxanthellae from the environment apparently occurs in the veliger stage of the clam's development (Fitt and Trench 1981).

When submerged, tridacnid clams gape and extrude their mantle folds. This action exposes the zooxanthellae in these tissues to sunlight and permits photosynthesis to take place in the zooxanthellae. Although the clams may receive some nutrition from digesting zooxanthellae, most of the metabolic gain from the symbiosis is believed to be derived from released photosynthates (Goreau et al. 1973, Trench et al. 1981), specifically glycerol and alanine (Muscatine 1967).

Both the zooxanthellae and the bivalve benefit from their relationship. The algae are sheltered, nourished with dissolved inorganic nutrients, and exposed to sunlight by their host. In turn, the clam has a ready source of inorganic and organic nutrients that are usually in low concentration in reef waters (Yonge 1936). Further, the uptake kinetics of inorganic compounds by this symbiosis seems to provide the capacity to take immediate advantage of spikes or surges in limiting nutrients, e.g., ammonium, much the same as has been shown in corals (Wilkerson and Trench 1986).

The bivalve host can rapidly incorporate carbohydrates produced by the zooxanthellae into its metabolic processes (Goreau et al. 1973). However, these carbohydrates are not a source of nitrogen (zooxanthellae are carbon fixers, not nitrogen fixers) and protein to the bivalve. To gain these food components, the bivalve must filter phytoplankton and microzooplankton from the surrounding water. This dependence on normal sources of bivalve nutrition in the tridacnids is evidenced by the presence of feeding gills, selective palps, and an elaborate digestive system (Yonge 1975). There are numerous observations of tridacnids kept in the dark with and without plankton. These experiments usually result in a loss of zooxanthellae and, if plankton is present, survival of the host.

Because of the symbiosis exhibited between zooxanthellae and tridacnid clams, a number of studies have focused on the role of the zooxanthellae in the bivalve host's nutrition. These studies have been limited to examining if and how nutrients are passed from the zooxanthellae to the bivalve, the role of light in controlling photosynthesis by the zooxanthellae, and determination of the production:respiration (P:R) ratio.

Early studies (see Goreau et al. 1973 and Fankboner and Reid 1990 for reviews) contended that zooxanthellae were captured by amebocytic blood cells, with digestion occurring in the blood stream and indigestible products being removed by the kidneys. Light and electron microscope studies by Trench et al. (1981) found no evidence for zooxanthellae in blood amebocytes. These authors also found no support for tridacnids digesting their symbionts and the undigested components collecting in the kidneys. In fact, the deposits in the kidneys were found to be mineral phosphorus. Using radioactively labeled carbon dioxide ($^{14}CO_2$), Goreau et al. (1973) were able to clearly show that uptake of label by the zooxanthellae and the eventual transmission of the labeled products via the blood to the rest of the bivalve. Thus, it appears that it is the photosynthetic products of the zooxanthellae that are utilized for food instead of the zooxanthellae themselves.

Tridacnids expose their zooxanthellae-containing mantle tissues only during the day and are nocturnally inactive (Reid et al. 1984). This observation also focuses on the importance of the zooxanthellae symbionts as a source of nutrition, because the great majority of plankton in coral reef waters is usually only over the reef during night, while the zooxanthellae require exposure to light.

Light plays a major role in the nutrition of tridacnid bivalves because of the light-limited nature of zooxanthellae. In essence, we must look at these clams as if they are primary producers rather than herbivores. Studies of oxygen uptake and release in *Tridacna maxima* (Trench et al. 1981) show that these bivalves are adapted to strong light intensities typical of shallow coral reef environments and show no signs of photoinhibition. By comparing light-driven evolution of oxygen (production

+ respiration = net production) to dark uptake of oxygen (respiration), Fisher et al. (1985) clearly showed P:R ratios greater than 1.0. At lower light intensities, smaller clams have higher P:R ratios, probably because more of their zooxanthellae are exposed to higher light intensities, while in larger clams, the thicker shells and bodies shade a sizable proportion of the zooxanthellae.

Trench et al. (1981) estimated the relative contribution of zooxanthellae photosynthesis to respiratory carbon demand by the clam *T. maxima*. They utilized direct observations of respiration, net photosynthesis, and the proportion of net carbon translocated from zooxanthellae to the clam following Porter's model (1980):

$$\%C = \frac{\left[\left(P_{net}\text{clam, 12hday}\right)\left(.375PQ^{-1}\right) + \left(\text{Rclam, 12hnight}\right)\left(.375RQ\right)\left(95\%\right)\right]\left(40\%\right)}{\left(\text{Rclam, 12hnight}\right)\left(.375RQ\right)\left(2\right)\left(95\%\right)}$$

where %C is carbon assimilated, P is production, PQ is photosynthetic quotient, R is respiration, and RQ is respiration quotient. From this model that only requires oxygen concentration observations, estimates of zooxanthellae contribution to clam respiratory carbon needs on sunny days were 77 to 84% and on cloudy days 62 to 63%. Thus, the photosynthetic carbon produced by the zooxanthellae may provide over half of the respiratory carbon requirements of these tridacnid clams. These authors stress, however, that due to the many assumptions made in the models, these estimates should not be considered definitive.

In more recent studies, Klumpp et al. (1992) and Hawkins and Klumpp (1995) found that giant clams deposit proportionally more carbon into their tissues relative to that respired than completely heterotrophic bivalves. Furthermore, small clams depended more on filter feeding (65%) for carbon needs, while large clams only acquire 34% of their carbon from this source — all of which supports the notion that with the development of clams from settlement to adult the symbionts also develop and become a major component in the clam's nutrition.

Another aspect of nutrition in this photoautotrophic symbiosis is that of the source and recycling of inorganic nutrients. In a series of studies, Fitt et al. (1993) and Belda et al. (1993) determined that nonsymbiotic larvae and young clams with few zooxanthellae release ammonium while large clams deplete ammonium and nitrate from sea water. Clams grown in elevated concentration of ammonium and phosphate showed an increase in the concentration of these elements in tissues of both the clam and the algae. Stoichiometric analysis revealed that the clams probably regulate the availability of phosphorus to the zooxanthellae. Using ^{15}N-ammonium, Hawkins and Klumpp (1995) found there was a net uptake of ammonium from sea water which was modulated by nutritional history, light history, ammonium concentration, and possible biological rhythmicity. Zooxanthellae also assimilated nitrogen excreted by host tissues. Most of the nitrogen released by the zooxanthellae was incorporated into the host tissues with no significant loss from the clam. They concluded that the photoautotrophic symbiosis of the giant clam confers a nutritional advantage over nonsymbiotic bivalves because this symbiotic relationship tightly recycles limiting nutrients in a nutrient poor environment.

Thus, tridacnid clams are nutritional opportunists in a nutrient-poor environment. They filter feed on suspended particles including diatoms, crustaceans, and zooxanthellae released from heat-stressed corals; take up dissolved organic matter; and incorporate photosynthates released by the zooxanthellae (Fankboner and Reid 1990). In doing so, they reflect essentially the same nutritional diversity as the dominant corals in their ecosystem.

Chemoautotrophic Symbioses

Chemoautotrophic symbioses are widespread in marine bivalve molluscs (Fisher 1990). Both chemoautotrophic or methanotrophic symbionts have been clearly demonstrated in numerous species of bivalves (Table 2). While the behavioral, physiological, and morphological adaptations to their symbionts vary among the different families of bivalves, there are many similarities. Unlike the photoautotrophic symbionts, most of the chemoautotrophic symbionts are intracellular, with the Thyasiridae an exception. The various bacterial symbionts appear to be associated with bacteriocytic cells and are often located in cellular vacuoles. These associations are found in a wide range of habitats, usually at the interface between aerobic and anaerobic environments. The successful bivalve symbioses appear to bridge this interface with numerous and often unique adaptations.

As the chemoautotrophic symbionts are relatively small (prokaryotic in size), a variety of methods have been used to demonstrate their existence. These approaches include electron microscopical identification, level of enzymatic activity, presence of lipopolysaccharide, stable isotopic ratios, presence of elemental sulfur, unique physiological adaptations, and the anatomy and habitat of the host (Fisher 1990). Enzymatic studies focus on the presence and activity of specific enzyme systems responsible for chemosynthesis. The enzymes of interest are:

1. Ribulose-1,5-bisphosphate carboxylase-oxygenase (RuBPC/O) catalyzes the first reaction in the Calvin Benson Cycle;
2. Adenosine-5′-phosphosulfate reductase (APS reductase) catalyzes the reaction of AMP and sulfite;
3. ATP sulfurylase catalyzes the reversible reaction of adenosine-phosphosulfate and pyrophosphate to yield ATP and sulfate;
4. Sulfate adenylyl transferase, in the same reaction as ATP sulfurylase, yields ADP instead;
5. Rhodanase catalyzes the splitting of thiosulfate into sulfite and sulfur;
6. Sulfide oxidase catalyzes the first step in the oxidation of hydrogen sulfide;
7. Nitrate reductase catalyzes the reduction of nitrate to nitrite;
8. Nitrate reductase catalyzes the reduction of nitrite to ammonia (see Fisher [1990] for a more detailed discussion).

Another group of tools in the ecologist's arsenal for identifying chemoautotrophic symbioses is the stable isotope ratios of carbon, nitrogen, and sulfur. Long known for their utility in determining the food source and web structure of ecological

TABLE 2
Bivalves with Chemoautotrophic Symbionts

Species	Visual EM	Enzymes	$\delta^{13}C$(%)	°S
Lucinidae				
Anodontia philippiana	+	Ru		
Codakia costata	+	Ru		
Codakia orbicularis		Ru, As, R	–23.2 to –28.3	
Codakia orbiculata		Ru		
Linga pennsylvanica		Ru		
Loripes lucinalis	+	Ru, Ar, As		
Lucina nassula		Ru	–23.0	
Lucina radians		Ru		
Lucinella divaricata	+	Ru, Ar, As		
Lucinoma aequizonata	+	Ru, As, R		+
Lucinoma annulata	+			
Lucinoma atlantis			–31.2 to –33.0	
Lucinoma borealis	+	Ru, Ar	–24.1 to –29	+
Myrtea spinifera	+	Ru, Ar, As	–23.1 to –24.2	+
Parvilucina multilineata	+	Ru		
Parvilucina tenuisculpta	+	Ru, As		+
Pseudomiltha floridana	+	Ru, Ar		+
Solemyidae				
Solemya reidi	+	Ru, As, Ar, R	–30	+
Solemya velesiana	+			
Solemya velum	+	Ru	–30.9 to –33.9	
Thyasiridae				
Thyasira equalis	+	Ru		
Thyasira flexuosa	+	Ru, Ar, As	–29.3	+
Thyasira gouldi	+		–32	
Thyasira sarsi	+	Ru, Ar, As	–28.2 to –31	+
Vesicomyidae				
Calyptogena elongata	+			+
Calyptogena laubieri	+			+
Calyptogena magnifica	+	Ru, As	–32.1 to –51.6	+
Calyptogena phaeseoliformis	+		–37.8 to –40.1	+
Calyptogena ponderosa	+		–31.2 to –39.1	+
Vesicomya cordata	+	Ru, As	–39.8	
Vesicomya gigas	+	Ru, As		
Mytilidae				
Bathymodiolus thermophilus	+	Ru, As	–30.5 to –37.1	

Note: Ar = APS reductase, As = ATP sulfurylase, R = ribulose 5′ kinase, Ru = ribulose bisphosphate carboxylase/oxygenase, °S = elemental sulfur in gills.

From Fisher 1990.

systems, these isotopic ratios are also useful in identifying chemoautotrophic symbioses. The sources of carbon, nitrogen, and sulfur elements for the various energy-generating reactions are quite different from those in nonsymbiotic bivalves. While the individual isotopic ratios of carbon ($\delta^{13}C$), nitrogen ($\delta^{15}N$), and sulfur ($d^{34}S$) can give evidence for food sources, and carbon and nitrogen isotopes can reflect trophic level, it is the broad range and use of multiple isotopic ratios that provides the most useful information both in symbiotic studies as well as in ecosystem level analyses.

Because elemental sulfur may serve as an energy storage product, a result from sulfide detoxification or as a by-product of sulfide oxidation, its presence in conjunction with other characteristics is a good indicator of chemoautotrophic symbioses. Although elemental sulfur is a good indicator, its presence is variable and is dependent on the physiological state of the symbiont (Fisher 1990).

The anatomy of the host, particularly its feeding structures, often suggests the availability of some type of nutritive symbioses. In bivalves, these characteristics include lack of a gut, hypertrophied gills, and reduced particulate feed ability. Also, if the host bivalve is limited to habitats where the substrates necessary for chemoautotrophiy are present, then chemoautotrophy might be suspected (Fisher 1990). All hosts for chemoautotrophic symbionts must provide their symbionts with an electron donor, an electron acceptor, and a carbon source. The electron donor is usually some form of reduced sulfur, either hydrogen sulfide or thiosulfate, the electron acceptor is oxygen, and the carbon source is carbon dioxide or bicarbonate. As bivalves primarily utilize aerobic metabolic pathways, and sulfide is toxic to most enzyme systems and disappears in the presence of oxygen, bivalves have developed complex mechanisms for providing their symbionts with both reduced sulfur and oxygen.

The bivalve species that have been suggested to contain chemoautotrophic symbionts and the evidence supporting these identifications are given in Table 2. In all species identified thus far, the symbionts are located in the gills and are intracellular in all but the Thyasiridae.

All members of the family Lucinidae that have been examined contain symbionts (Fisher 1990). These bivalves typically have a reduced gut, reduced labial palps, no incurrent siphon, and large fleshy gills (Allen 1958). The Lucinidae are thought to obtain sulfide from the reduced sediments surrounding their burrows and oxygen from the overlaying water.

Of the several chemoautotrophic species in the Mytilidae, *Bathymodiolus thermophilus*, the deep sea hydrothermal vent mussel, is the best studied (Fisher 1990, Kenk and Wilson 1985). These mussels have been found in a wide range of environmental conditions in the area of the hydrothermal vents, and this distribution has been attributed to their ability to obtain nutrition from symbioses (Fisher 1990), filter feeding (Le Pennec and Prieur 1984) and dissolved organic material (Fiala-Medioni et al. 1986). Stable isotope ratios of mussels taken near the Rose Garden hydrothermal vent site on the Galapagos rift support the suggestion that multiple sources of nutrition are important to these animals (Fisher 1990).

The family Solemyidae was initially an enigma to science because it is composed of relatively large prosobranch bivalves with reduced or absent digestive tract and feeding abilities (Reid 1980). All members of this family are now thought to contain chemoautotrophic symbionts. Further, these shallow water estuarine clams typically make deep burrows in organically rich sediments that penetrate both oxidized and reduced sediment layers. Dense populations of *Solemya reidi* are found in sites polluted from paper mill effluent and sewage sludge deposits that are high in reduced sulfides. Despite their symbiotic nature and their lack of a gut, these clams reach a size of up to 5 cm.

The Thyasiridae are closely related to the Lucinidae, but they are notable in that their symbiotic bacteria are extracellular. The extracellular nature of the symbionts has been interpreted as indicating an early stage in the development of chemoautotrophic symbioses in this group compared to those groups with intracellular symbionts (Fisher 1990).

Vesicomyids have been found in a variety of middepth to deep sea sulfide-containing habitats, including hydrothermal vents, subduction zones, saline seeps, hydrocarbon seeps, and deep sea reducing sediments. Some species of these clams may burrow through the sediments, while others are epifaunal. The vent clam *Calyptogena magnifica* is one of the better studied of these chemoautotrophs. Here again, it is evident that the clam utilizes the interface between the oxidized and reduced environments to the best advantage of the symbiosis. These epifaunal clams of the hydrothermal vents orient their bodies such that the foot is exposed to warm reducing vent waters rich in hydrogen sulfide, while cooler overlaying water high in oxygen is pumped through the clam.

Thus, a small number of bacterial species have succeeded in forming an endosymbiotic relationship with marine bivalves. As the bacteria are mainly sulfide oxidizers, these microbes benefit from the dependable interface (Reid and Brand 1986) the bivalve provides between the aerobic sediment surface and water column and the anaerobic environments of reducing sediments and hydrothermal vent environments. The bivalves benefit from an additional source of food or energy, and this feature may allow them to exist in environments they might otherwise be unable to inhabit.

PRODUCTION (P)

Production is defined as the accretion of somatic tissues and/or gametes and has been extensively studied in bivalves (Hawkins and Bayne 1992). Production includes not only material and energy stored as new tissues and represented as growth of the individual (P_g), but also material that goes into gametes (P_r). Hawkins and Bayne (1992) have stated that production is controlled by the interaction between endogenous processes that pertain to the organism and exogenous factors from the external environment. Exogenous influences include quality and quantity of food, temperature, and salinity, and have been discussed elsewhere in this text. The major endogenous factors are body size, physiological status, and genotype. These internal influences determine maximal capacity for production.

Production Due to Growth (P_g)

Growth is one of the most-used measures of an organism's vitality in a given environment. In bivalves, the animal's size is directly related to its age, and this cumulative increase in biomass with respect to time is termed "absolute growth," while the percentage increase in biomass per unit time is "relative growth" (Seed 1976).

As growth represents changes in bivalve size, it is most often measured as shell length, weight, and volume. There may be complications in just measuring single shell parameters, as shell growth may be different from soft body growth due to environmental factors or variations in the reproductive cycle of the bivalve. It is the soft body that carries out the living processes of the animal, not the shell. Thus, allometric relationships are often developed between shell parameters and body weight in order to nondestructively estimate soft body biomass on living bivalves (Dame 1972).

Growth measurement

Some methods used to assess growth in bivalves are following size changes in marked individuals, tracking size frequency distributions, measuring shell growth rings in relation to biomass, and observing radioactive tracer uptake. To determine seasonal growth rates as well as the influence of size and local environmental conditions, marked bivalves are often kept in wire cages at a particular depth or intertidal elevation. The cages partially protect the bivalve from predation and make them easier to retrieve than individual animals. Generally, cages do not influence growth, but Harger (1970) found that mussels kept in intertidal cages grew faster than those not in cages, possibly because wave action was reduced within the cages.

Following a particular cohort by measuring size frequency distribution through time is a common technique among population ecologists (see Chapter 4). In this method, a given year class is followed, and the change in the average size of the mode is equivalent to average growth. This method only works well in bivalve species with a very narrow reproductive or recruitment period (Craig and Oertel 1966). If the recruitment period is extended, i.e., *Crassostrea virginica* (Dame 1972), or growth rates are variable, i.e., *Mytilus* (Seed 1976), then the method has limited application.

Shell rings or growth lines are defined as abrupt or repetitive changes in the character of accreting bivalve tissue (Dillon and Clark 1980) and, depending on the circumstances, may be a post mortum indicator of growth. These lines are evident in some bivalves when they experience well-defined seasonal conditions, particularly suspended periods of shell growth (Seed 1976). Such periods are usually associated with extremes of temperature, wave action, or reproductive periods. Bivalves from environments with uniform conditions generally do not exhibit rings, and line formation is variable both between habitats and with a habitat. Thus, care should be taken to check the reliability of rings in each locality.

Shell growth

Growth of shell is quite different from growth of somatic and reproductive tissues. The bivalve shell is composed of both an organic matrix and crystals of calcium carbonate ($CaCO_3$). There are three major aspects of shell growth and formation: (1) the role of metabolism in $CaCO_3$ formation and the synthesis of the organic matrix; (2) secretion of shell components by the mantle; and (3) $CaCO_3$ crystal growth (Wilbur 1964; Wilbur and Saleuddin 1983).

As $CaCO_3$ and CO_2 are both components of the calcium carbonate–carbon dioxide buffer system so prevalent in sea water, changes in the concentrations of any component in this buffer system influence the amount of the other components.

$$CO_2 + H_2O \leftrightarrow H_2CO_3 \leftrightarrow HCO_3^- + H^+ \leftrightarrow CO_3^{2-} + H^+ \leftrightarrow CaCO_3 \leftrightarrow Ca^{2+}$$

$$\text{alkaline} \longrightarrow \qquad\qquad\qquad\qquad \longleftarrow \text{acid}$$

Increasingly alkaline conditions will drive this buffer system to the right with an increase in $CaCO_3$, assuming sufficient calcium ions are present, and increasing acidity will drive the system to the left with the dissolving of $CaCO_3$ and eventual production of CO_2. The CO_2 concentration within the bivalve's tissues can also be a function of intermediary metabolism, and combined with the buffer system, potentially controls shell crystal formation (Wilbur 1964). The amount of calcium deposited in shell is also a function of the calcium concentration in the surrounding water and time of submergence. Early studies of intertidal bivalves clearly supported the above processes in that the longer the animal is submerged the greater the calcium uptake and consequently shell growth is higher (Rao 1953; Baird and Drinnan 1956).

The organic matrix is secreted in insoluble form by the mantle and deposited as a layer on the inner shell surface. The matrix is proteinaceous and is the substrate upon which $CaCO_3$ crystals initiate and grow.

While measuring shell growth by weight in water (Havinga 1928; Andrews 1961) or by linear measures is straightforward, determining the organic matrix component of the shell, a relatively small component of growth, as a percentage of total weight is difficult. Although small in mass, in terms of calories or organic carbon, the organic matrix may be an important component of growth. The two most common ways of estimating the biomass of the organic matrix are through careful ashing in a furnace or acid dissolution of shell and weighing the remaining dried organic material (Dame 1972; Price et al. 1970).

In addition to seasonal and reproductive shell growth rings that are visually external, microscopic growth increments or lines ranging from 1 to 100 μm thick are discernible within bivalve shells (Rhoads and Pannella 1970; Lutz and Rhoads 1980). These deposition increments are laid down during each 24-hour period of physiological activity (House and Farrow 1968; Clark 1968; Clark 1974). It is generally accepted that many, if not all, microgrowth increments in bivalve shells are a reflection of variations in the relative proportions of organic material and calcium carbonate (Lutz and Rhoads 1980). In many bivalves, these growth lines appear to be correlated with the tides and the fact that the valves are open during submergence and closed during exposure. To explain these lines, Lutz and Rhoads (1977) proposed that differences in organic concentrations in the various regions of the shell were simply residues left behind as a result of dissolution of shell material during periods of anaerobiosis. Subsequent studies have supported their hypothesis

(Gordon and Carriker 1978; Lutz and Rhoads 1980). Microgrowth increments have been documented in five temporal categories: (1) semidiurnal and diurnal, (2) fortnightly, (3) monthly, (4) annual, and (5) semiperiodic or random events (Lutz and Rhoads 1980). These microscopic lines provide information about environmental and physiological conditions of individual animals from both current and past environments (Table 3). Thus, the bivalve shell may be considered a long-term continuous environmental recorder.

TABLE 3
Physiological and Environmental Events Recorded in Molluscan Growth Patterns

Events	Characteristics	Time and Place	Ecological Application
Circadian Rhythms	Dark and light shell layers	Daily	Physiology
Spawning	Sudden depositional break	Variable by species and place	Reproductive period and duration
Winter	Gradual reduction in deposition rate	Shallow water at mid and high latitudes	Climate, latitude, and geography
Summer	Thickest daily increment		Climate, latitude, and geography
Tides	Regular recurrent patterns with 15- to 30-d cycle	Intertidal and shallow water	Sea level change
Storms	Sudden depositional break with rapid recovery	Storm frequency	Weather

From Rhoads and Pannella 1970.

To summarize, shell growth is a function of calcium availability in water and, therefore, of the amount of water pump across the bivalve's tissues, the pH of the bivalve's internal system, the intermediary metabolism of the animal, and the environmental temperature that controls the rates of most of these processes. Although shell size is related to soft body tissue mass, it is not a direct measure of body size, because some of the controlling factors are different.

Production Due to Reproduction (P_r)

In addition to the energy and materials bivalves allocate to shell and body growth, they place considerable amounts of production into gonadal tissues (P_r). These tissues develop into gametes that are lost from the individual during spawning. Consequently, there is a cyclical storage and loss of reproductive products that is controlled by endogenous and exogenous factors.

There are numerous methods used to estimate the amount of production allocated to reproduction. The amount of P_r can be directly estimated by inducing the bivalve to spawn in the laboratory and then counting or weighing the gametes released.

Indirectly, P_r can be estimated from allometric relationships relating weight loss due to spawning to body weight or shell size (Bayne and Worrall 1980) and by estimating the proportion of gonad lost from the proportion of spatfall in a given time period (Dame 1976).

The proportion of production devoted to reproduction is low in young bivalves, but gradually increases to a dominant proportion in adults. In the largest mussels, reproductive tissues may account for more than 50% of the soft body weight, and P_r may be >90% of total production (P) (Seed and Suchanek 1992; Thompson 1979). Bivalves may "conservatively" utilize nutrient reserves to buffer gametogenesis from environmental changes (Hawkins et al. 1985). These reserves may come from somatic tissues. Generally, gametogenesis is "opportunistic," being supported by feeding alone (Hawkins et al. 1985).

The condition index for bivalves relates the proportion of the shell cavity that is occupied by soft body tissue. Such indices have been used to follow seasonal change in gross nutrient reserves or meat quality, as well as to monitor various pollutants (see Chapter 8) and diseases (Crosby and Gale 1990). In a comparison of various methods of determining the condition index, Crosby and Gale concluded that the gravimetric formulation

$$\text{Condition Index} = \left[\frac{\text{dry soft tissue wt(g)} \times 1000}{\text{internal shell cavity capacity (g)}}\right]$$

has less measuring errors, lower coefficients of variation, is the easiest and fastest to use, and is most meaningful as an index of current bivalve nutritive status and recent stress. These investigators stress the need for further investigations into the seasonal and age-dependent relationships between tissue/shell components and gravimetric condition indices in order to gain new insights into bivalve energetics.

Growth Models

Numerous models have been developed to describe the growth of individual organisms. The most useful models are those in which the components and constants have biological meaning and a few of these will be reviewed here.

The simplest formulation is the monomolecular model

$$l_t = L_\infty\left(1 - e^{kt}\right)$$

where l_t is the size at time t; L_∞ is the maximum size; and k is a constant representing the rate at which maximum size is approached. In this model there is no inflection point, and growth rate declines linearly with increasing size. Maximum size can be determined from observed data or from a Walford plot (Walford 1946):

$$l_t = bL_{t+1} + a$$

Where this linear relationship intercepts the no-size-change line is an estimate of maximum size. This function is often used to represent the later portions of life (Richards 1959).

The autocatalytic or logistic model

$$l_t = \frac{L_\infty}{\left(1 - e^{-kt}\right)}$$

produces a symmetrical or s-shaped curve about an inflection point. Its relative growth rate declines linearly with increasing size. This model has been used as a convenient empirical curve (Richards 1959).

The Gompertz model

$$l_t = L_\infty e^{-e^{-kt}}$$

or

$$\log l_t = \log L_\infty \left[1 - e^{-kt}\right]$$

resembles the logistic model, but is asymmetrical. Linear relations exist between its relative growth rate and log l and between log l and time.

The von Bertalanffy model (Bertalanffy 1957):

$$l_t = L_\infty \left[1 - e^{-k(t-t_0)}\right]$$

in its general form includes all of the preceding models. This model has been favored by biologists because of this generality and its derivation from allometric and metabolic relationships with biological meaning (Figure 20).

Several investigators (Taylor 1959, 1960; Ursin 1963) have adapted the von Bertalanffy model to temperature and bivalve growth. These investigators use the product of the number of days and the temperature for those days, day-degrees, to represent the environmental variable. The following is an elegant example of utilizing day-degrees to represent temperature influences on growth:

$$l_D = L_\infty \left[1 - e^{-k_D(D-D_0)}\right]$$

where l_D is length at D day-degrees, K_D is the rate constant determined as k/D_y (D_y = annual sum of day-degrees) and $D_0 = t_0$ (see Bayne and Worrall [1980] for an example).

The preceding models utilize size data to estimate growth. An alternative approach to estimating growth is based on the energy budget equation and generating

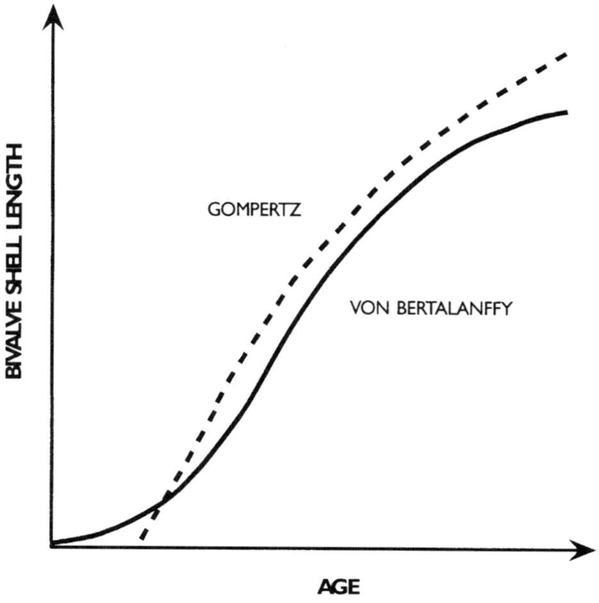

FIGURE 20. A generalized comparison of the Gompertz and Von Bertalanffy growth curves in bivalves.

an estimate of growth potential or scope for growth (Warren and Davis 1967; Bayne et al. 1976). When these observations are integrated over time and applied to individuals of different sizes, an average growth curve can be derived.

Respiration (R)

At the ecosystem level, respiration represents a measure of the energetic cost of supporting a given bivalve. Respiration is a unique component of the energy budget because it is entirely a loss term. The matter associated with respiration is recycled as CO_2 and water and energy is lost as waste heat to the environment. At the level of the organism, respiration represents all the work a bivalve must perform in order to survive. Finally, respiration is the most thoroughly and precisely measured process in the energy budget. Because of this, there is an extensive body of knowledge and theory explaining respiration.

In bivalves, respiratory gas exchange between the water and the animal occurs primarily at the gill. The bivalve gill has a high surface area due to the folding of the tissues or demibranchs, and this feature increases the ability of the animal to exchange respiratory gases. The demibranchs are also well supplied with blood vessels and this characteristic ensures the rapid movement of gases into and out of the bivalve. Last, large volumes of water are pumped or moved by the cilia on the gill. This action provides a constant renewal of water within the bivalve so that fresh water with higher oxygen levels is constantly provided and stale waters with higher waste gas concentrations are constantly flushed from the animal. Thus, in addition to being the main feeding organ of filter feeders, the gill is also the main respiratory organ.

Aerobic respiration involves the utilization of organic carbon compounds as sources of matter and energy. In its most fundamental form:

$$\text{Organic-C} + O_2 \rightarrow CO_2 + H_2O + \text{Energy}$$

Estimates of metabolism usually focus on the measurement of changes in one of the components of the preceding relationship. In actuality, oxygen concentration is the most commonly measured estimator, although carbon dioxide and heat are also used. Once one component of the metabolic equation is known, the others can be calculated. If the organic carbon source is unknown or varying, then measurement of both oxygen and carbon dioxide allows the computation of the repiratory quotient (RQ) and thus an estimate of the type of carbon compounds being metabolized.

Oxygen consumption is directly related to the size of the bivalve by the allometric power function:

$$R = aW^b$$

where R is oxygen uptake, W is the body weight of the bivalve, a is the intercept, and b is the slope. Surveys of respiration-to-weight relationships for many different animals, including bivalves (Hemmingsen 1960) found that the coefficient b is about 0.75. In bivalves, estimates of b may vary from 0.5 to 1.0, but Bayne et al. (1976) note that deviations from $b = 0.75$ are often attributed to a narrow size range in the observed animals. Kruger (1960) recommends a size range of 1:50 from smallest to largest individuals to provide reliable estimates of a and b.

Peters (1983) has made a number of points concerning the oxygen consumption to size relationship. First, the apparent constancy of the slope ($b = 0.75$) shows that the proportionality is unaffected by size. Second, the positive value of b means larger animals respire at higher rates than smaller individuals. Third, as $b < 1$ the metabolic rate rises more slowly than body size. Finally, the data are scattered, which suggests the relationship should not be interpreted as deterministic.

As metabolism is the summed energetic cost of an organism's life processes, the relationship between size and respiration should be reflected in other biological relationships. Thus, the power function, $b = 3/4$, is often successful in describing scaling for other rates. This parallelism is called the principle of similitude (Thompson 1961). Although the surface law and dimensional analysis are most often given as the explanation for this allometric relationship, each has its problems (see Peters 1983) and there is no consensus.

Because most adult bivalves are relatively sessile in their benthic habitats, many of these animals are exposed to anaerobic as well as aerobic conditions. Hence, bivalves are defined as facultative anaerobes, i.e., they can live aerobically or anaerobically, but preferentially use oxygen for respiration because of its higher efficiency in producing energy during intermediary metabolism (Zwaan and Mathieu 1992). Anaerobic conditions can occur in bivalves when they seal their valves during exposure at low tide or when their external environment has reduced oxygen concentration. In some cases, hydrothermal vents and some sediments, the bivalves live at the interface between aerobic and anaerobic environments and through the presence of symbiotic bacteria utilize inorganic compounds as energy sources. Some of these compounds, i.e., H_2S, are effective poisons to aerobic respiration by inhibiting cytochrome function and blocking O_2 uptake by hemoglobin.

Anaerobic metabolism is a normal part of the production of energy by metazoans. However, energy production through anaerobic metabolism is somewhat lower than through aerobic metabolism. In adult bivalves lacking red blood cells, the animal can switch between aerobic and anaerobic metabolism, depending on oxygen concentrations in the environment (Mangum and Van Winkle 1973). There is an ecological advantage to switching to anaerobic metabolism. Anaerobic metabolism is reduced and this reduction lowers total metabolic activity and energy demand. In intertidal bivalves, feeding stops during low tide exposure; therefore, the carbon sources for metabolism are reduced and lower metabolism (anaerobic) is favored. Reduced metabolism and anaerobiosis during winter months are thought to be important factors in freezing temperature survival in *Geukensia* (Aarset 1982). In this case, anaerobic metabolism end products may act as antifreeze or cryoprotectant (Loomis et al. 1988).

While the determination of aerobic respiration through oxygen uptake is well known and developed, determination of anaerobic metabolism and total metabolism (aerobic + anaerobic) are less so. Anaerobic metabolism is often estimated indirectly by measuring anaerobic metabolic endproducts. Total metabolism (Figure 21) can

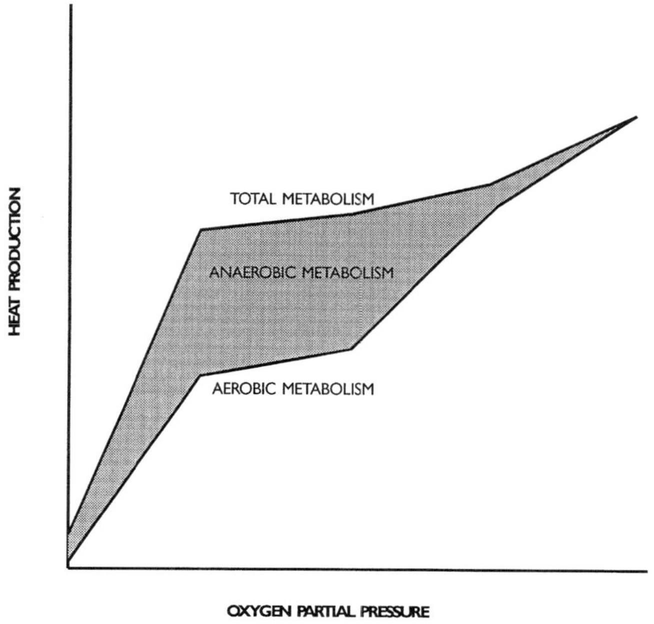

FIGURE 21. The general relationship between heat production, oxygen partial pressure, and aerobic, anaerobic, and total metabolism in bivalves.

be estimated by summing aerobic and anaerobic estimates or by directly measuring total heat production with a calorimeter (Shick et al. 1983). The heat production method was first used on invertebrates by Gnaiger (1977) and later applied to aerobic bivalves by Hammen (1979) and anaerobic bivalves by Hammen (1980) and Pamatmat (1979). The indirect approach to estimating anaerobic metabolism from endproducts in bivalves was led by Zwaan (1977) and others. Originally, it was thought that the direct production method produced a greater estimate of metabolism than could be accounted for by endproducts (Gnaiger 1980). However, later studies (Shick et al. 1983) showed that individual organism variability was large enough to explain the discrepancy. Thus, the indirect and direct methods of determining metabolism are comparable.

As noted earlier, oxygen consumption increases with bivalve size according to the power law (b = 0.7). The relationship between anaerobic heat production rate and bivalve body size is different with a b = 1.0 (Pamatmat 1979). In general, anaerobic metabolism for intertidal mussels is about 7.5% of aerobic metabolism (Pamatmat 1979).

Excretion

Marine bivalves produce excretory waste products as a result of incomplete digestion and metabolism. Thus, various forms of carbon, nitrogen, phosphorus, and silicon can be found in the waste of bivalves. Three of these components, nitrogen, phosphorus, and silicon, are often limiting in marine environments and, for this reason, it is important that the flows and processes involving these constituents be understood (see Chapter 6).

While metabolic degradation of organic compounds takes place in all living tissues, the excretion of the end products of metabolism usually takes place via the kidneys and across tissue surfaces, particularly the gills. Both of these locations have high surface-to-volume ratios that allow maximum exchange across their surfaces. It has long been thought that filtration of the blood occurs in the pericardial coelom of bivalves, and the products of this filtration flow to the kidney where secretion and reabsorption take place (Martin and Harrison 1964).

Bivalve molluscs excrete nitrogen as a major endproduct of protein and nucleic acid metabolism. In studies using incubation chambers, ammonium is the major excretory product with urea, uric acid, purines, and free amino acids (FAA) in lesser amounts (Hammen 1980). However, recent studies using high performance liquid chromatography (HPLC), a technique that directly measures the net fluxes of metabolites during a single passage of natural sea water over a bivalve's gill, have shown that ammonia is the only nitrogenous waste product excreted by some bivalves (Hawkins and Bayne 1992).

As noted above, incubation techniques indicate that ammonia is the most common nitrogenous excretory product, varying from 37 to 78% of total nitrogen excreted by bivalves (Table 4). Some of the ammonia produced within the bivalves' tissues may be utilized to synthesize organic products, but there is always a surplus

TABLE 4
Representative Percent of Total Nitrogen of Specific Excretion Products By Selected Marine Bivalves

	%N				
Species	**NH_4**	**Urea**	**FAA**	**Uric Acid**	**Ref.**
Crassostrea virginica	68	8	21	3	Hammen 1980
Donax serra	77	2	21		Cockroff 1990
Donax sordidus	78	1	22		Cockroff 1990
Mercenaria mercenaria	66		30	4	Hammen 1980
Geukensia demissus	64		36		Hammen 1980
Mytilus edulis	67–41	4	55		Bayne and Scullard 1977
Mytilus galoprovincialis	58–37		42–63		Bayne and Scullard 1977
Solemya velum	70		27	3	Hammen 1980
Tagelus plebius	50		31	19	Hammen 1980
Teredo navalis	55		10	35	Hammen 1980

that is excreted to the external environment. It is unclear at this time whether the FAAs observed in incubation studies are the actual products of bivalves or other heterotrophs in the water of the chamber. At the ecosystem level of investigation, FAAs are often referred to as DON, or dissolved organic nitrogen.

Bayne et al. (1976) note that the rates of nitrogen excretion are extremely variable, and this variability is probably due to marked seasonal changes in nutrient storage and utilization of reserves. There is evidence that organism size and feeding status together with environmental temperature, salinity, and exposure to air can all influence the rate of nitrogen excretion.

As with oxygen consumption, nitrogen excretion increases with bivalve size according to the following equation:

$$N = aW^b$$

where N is the rate of nitrogen excretion, W is the weight of the body and a and b are coefficients (Hammen 1980). Normally, during starvation bivalves metabolize stored glycogen, but as these reserves are used up, they are forced to increase protein catabolism as a source of energy (Bayne et al. 1976). In *Mytilus edulis*, ammonia excretion rises with temperature, but there is no acclimation (Bayne and Scullard 1977). Numerous investigators (see Bayne et al. 1976) have found that the rate of ammonia excretion increases as salinity decreases. These observations are supported by a review by Griffiths and Griffiths (1987) that found that the coefficients of the excretion–size relationship are strongly influenced by size, temperature, salinity, and exposure.

REFERENCES

Aarset, A.V. 1982. Freezing tolerance in intertidal invertebrates, *Comp. Biochem. Physiol.*, 73A, 571–580.

Allen, J.A. 1958. On the basic form and adaptations to habitat in the Lucinacea, *Phil. Trans. Roy. Soc. Ser. B*, 241, 421–484.

Allen, J.A. and Sanders, H.L. 1966. Adaptations to abyssal life as shown by the bivalve *Abra profundorum, Deep Sea Res.*, 13, 1175–1184.

Andrews, J.D. 1961. Measurement of shell growth in oysters by weighing in water, *Proc. Natl. Shellfish. Assoc. Add.*, 1–11.

Atkins, D. 1938. On the ciliary mechanisms and interrelationships of lamellibranchs. VII. Laterofrontal cilia of the gill filaments and their phylogenetic value, *Q. J. Microsc. Sci.*, 80, 345–436.

Baird, R.H. and Drinnan, R.E. 1956. The ratio of shell to meat in *Mytilus* as a function of tidal exposure to air, *J. Cons. Int. Explor. Mer*, 22, 329–336.

Bayne, B.L., Hawkins, A.J.S., and Navarro, E. 1988. Feeding and digestion in suspension-feeding bivalve molluscs: the relevance of physiological compensations, *Am. Zool.*, 28, 147–159.

Bayne, B.L. and Newell, R.C. 1983. Physiological energetics of marine molluscs, in Saleuddin, A.S.M. and Wilbur, K.M., Eds., *The Mollusca*, Vol. 4, Academic Press, New York, 407–515.

Bayne, B.L. and Scullard, C. 1977. Rates of nitrogen excretion by species of *Mytilus, J. Mar. Biol. Assoc. U.K.*, 57, 355–369.

Bayne, B.L., Thompson, R.J., and Widdows, J. 1976. Physiology I., in Bayne, B.L., Ed., *Marine Mussels: Their Ecology and Physiology*, Cambridge University Press, New York, 121–206.

Bayne, B.L. and Worrall, C.M. 1980. Growth and production of mussels (*Mytilus edulis*) from two populations, *Mar. Ecol. Prog. Ser.*, 3, 317–328.

Belda, C.A., Lucas, J.S., and Yellowlees, D. 1993. Nutrient limitation in the giant clam — zooxanthellae symbiosis: effects of nutrient supplements on growth of the symbiotic partners, *Mar. Biol.*, 117, 655–664.

Bender, K., and Davis, W.R. 1984. The effects of feeding by *Yoldia limatula* on bioturbation, *Ophelia*, 23, 91–100.

Bertalanffy, L. von. 1957. Quantitative laws in metabolism and growth, *Q. Rev. Biol.*, 32, 215–231.

Birkbeck, T.H. and McHenery, J.G. 1982. Degradation of bacteria by *Mytilus edulis*, *Mar. Biol.*, 72, 7–15.

Buxton, C.D., Newell, R.C., and Field, J.G. 1981. Response-surface analysis of the combined effects of exposure and acclimation temperatures on filtration, oxygen consumption and scope for growth in the oyster *Ostrea edulis*, *Mar. Ecol. Prog. Ser.*, 6, 73–82.

Calow, P. 1975. The feeding strategies of two freshwater gastropods, *Ancylus fluviatilis* and *Planorbis contortus,* with a comparison of field and laboratory estimates of food absorption rate, *Oecologia*, 20, 51–63.

Calow, P. and Fletcher, C.R. 1972. A new radiotracer technique involving ^{14}C and ^{51}Cr for estimating the assimilation efficiency of aquatic primary producers, *Oecologia*, 9, 155–170.

Clark, G.R. 1968. Mollusc shell: daily growth lines, *Science*, 161, 800–803.

Clark, G.R. 1974. Growth lines in invertebrate skeletons, *Annu. Rev. Earth Planet. Sci.*, 2, 77–99.

Cockcroft, A.C. 1990. Nitrogen excretion by the surf zone bivalves *Donax serra* and *D. Sordidus*, *Mar. Ecol. Prog. Ser.*, 60, 57–65.

Coughlan, J. 1969. The estimation of filtering rate from the clearance of suspensions, *Mar. Biol.*, 2, 356–358.

Craig, G.Y. and Oertel, G. 1966. Deterministic model of living and fossil populations of animals, *Q. J. Geol. Soc. London*, 122, 315–355.

Crisp, D.J. 1984. Energy flow measurements, in Holme, N.A. and McIntyre, A.D., Eds., *Methods for the Study of Marine Benthos*, IBP Handbook 16, Blackwell Scientific, Oxford, 285–369.

Crosby, M. P. 1987. Utilization of Detrital Complexes by the Oyster *Crassostrea virginica* (Gmelin), Ph. D. thesis, University of Maryland, College Park.

Crosby, M.P. and Gale, L.D. 1990. A review and evaluation of bivalve condition index methodologies with a suggested standard method, *J. Shellfish Res.*, 9, 233–237.

Crosby, M.P., Langdon, C.J., and Newell, R.I.E. 1989. Importance of refractory plant material to the carbon budget of the oyster *Crassostrea virginica*, *Mar. Biol.*, 100, 343–352.

Crosby, M.P. and Newell, R.I.E. 1990. Bacterial mediation in the utilization of carbon and nitrogen from detrital complexes by *Crassostrea virginica*, *Limnol. Oceanogr.*, 35, 625–639.

Crosby, N.D. and Reid, R.G.B. 1971. Relationships between food, phylogeny, and cellulose digestion in the bivalvia, *Can. J. Zool.*, 49, 617–622.

Dame, R.F. 1972. The ecological energies of growth, respiration and assimilation in the intertidal American oyster, *Crassostrea virginica*, *Mar. Biol.*, 17, 243–250.

Dame, R.F. 1976. Energy flow in an intertidal oyster population, *Estuarine Coastal Mar. Sci.*, 4, 243–253.

Dillon, J.F. and Clark, G.R. 1980. Growth-line analysis as a test for contemporaneity in populations, in Rhoads, D.C. and Lutz, R.A., Eds., *Skeletal Growth of Aquatic Organisms*, Plenum Press, New York, 395–415.

Doering, P.H. and Oviatt, D.A. 1986. Application of filtration rate models to field populations of bivalves: an assessment using experimental mesocosms, *Mar. Ecol. Prog. Ser.*, 31, 265–275.

Fankboner, P.V. and Reid, R.G.B. 1990. Nutrition in giant clams (Tridacnidae), in Morton, B., Ed., *The Bivalvia — Proc. Mem. Symp. in Honour of Sir Charles Maurice Yonge, Edinburgh, 1986*, Hong Kong University Press, Hong Kong, 195–209.

Fiala-Medioni, A., Alayse, A.M., and Cachet, G. 1986. Evidence of *in situ* uptake and incorporation of bicarbonate and amino acids by the hydrothermal vent mussel, *J. Exp. Mar. Biol. Ecol.*, 96, 191–198.

Fisher, C.R. 1990. Chemoautotrophic and methanotrophic symbioses in marine invertebrates, *Rev. Aquat. Sci.*, 2, 399–436.

Fisher, C.R., Fitt, W.K., and Trench, R.K. 1985. Photosynthesis and respiration in *Tridacna gigas* as a function of irradiance and size, *Biol. Bull.*, 169, 230–245.

Fitt, W.K., Rees, T.A.V., Braley, R.D., Lucas, J.S., and Yellowlees, D. 1993. Nitrogen flux in giant clams: size-dependency and relationship to zooxanthellae density and clam biomass in the uptake of dissolved inorganic nitrogen, *Mar. Biol.*, 117, 381–386.

Fitt, W.K. and Trench, R.K. 1981. Spawning, development, and acquisition of zooxanthellae by *Tridacna squamosa*, *Biol. Bull.*, 161, 213–235.

Gnaiger, E. 1977. Thermodynamic considerations of invertebrate anoxibiosis, in Lamprecht, I. and Schaarschmidt, B., Eds., *Applications of Calorimetry in Life Sciences*, Walter de Gruyter, Berlin, 281–303.

Gnaiger, E. 1980. Energetics of invertebrate anoxibiosis: direct calorimetry in aquatic oligochaetes, *FEBS Letts.*, 112, 239–242.

Goreau, T.F., Goreau, N.I., and Yonge, C.M. 1973. On the utilization of photosynthetic products from zooxanthellae and of a dissolved amino acid in *Tridacna maxima, J. Zool. Lond.*, 169, 417–454.

Gordon, J. and Carriker, M.R. 1978. Growth lines in bivalve mollusk: subdaily patterns and dissolution of the shell, *Science*, 202, 519–521.

Grave, C. 1916. The process of feeding in the oyster, *Science*, 44, 178–181.

Griffiths, C.L. and Griffiths, R.J. 1987. Bivalvia, in Pandian, T.J. and Vernberg, F.J., Eds., *Animal Energetics*, Academic Press, New York, 1–88.

Hammen, C.S. 1979. Metabolic rates of marine bivalve molluscs determined by calorimetry, *Comp. Biochem. Physiol.*, 62A, 955–959.

Hammen, C.S. 1980. Total energy metabolism of marine bivalve mollusks in anaerobic and aerobic states, *Comp. Biochem. Physiol.*, 67A, 617–621.

Hammen, C.S. 1980. *Marine Invertebrates: Comparative Physiology*, University Press of New England, Hanover, NH, 127 pp.

Harger, J.R.E. 1970. The effect of wave impact on some aspects of the biology of sea mussels, *Veliger*, 12, 401–414.

Haven, D.S. and Morales-Alamo, R. 1970. Filtration of particles from suspension by the American oyster *Crassostrea virginica*, *Biol. Bull.*, 139, 248–264.

Havinga, B. 1928. The daily rate of growth of oysters during summer, *J. Cons.*, 3, 231–245.

Hawkins, A.J.S. and Bayne, B.L. 1992. Physiological interrelations, and the regulation of production, in Gosling, E., Ed., *The Mussel Mytilus: Ecology, Physiology, Genetics and Culture*, Elsevier, Amsterdam, 171–222.

Hawkins, A.J.S. and Klumpp, D.W. 1995. Nutrition of the giant clam *Tridacna gigas* (L.). II. Relative contributions of filter-feeding and ammonium-nitrogen acquired and recycled by symbiotic alga toward total nitrogen requirements for tissue growth and metabolism, *J. Exp. Mar. Biol. Ecol.*, 190, 263–290.

Hawkins, A.J.S., Salkeld, P.N., Bayne, B.L., Gnaiger, E., and Lowe, D.M. 1985. Feeding and resource allocation of the mussel *Mytilus edulis*: evidence for time-averaged optimization, *Mar. Ecol. Prog. Ser.*, 20, 273–287.

Hemmingsen, A.M. 1960. Energy metabolism as related to body size and respiratory surfaces and its evolution, *Rep. Steno Mem. Hosp.*, 4, 7–58.

Hildreth, D.I. and Crisp, D.J. 1976. A corrected formula for calculation of filtration rate of bivalve molluscs in an experimental flowing system, *J. Mar. Biol. Assoc. U.K.*, 56, 111–120.

House, M.R. and Farrow, G.E. 1968. Daily growth banding in the shell of the cockle *Cardium edule, Nature*, 219, 1384–1386.

Hughes, R.N. 1969. A study of feeding in *Scrobicularia plana*, *J. Mar. Biol. Assoc. U.K.*, 49, 805–823.

Hughes, R.N. 1980. Optimal foraging in the marine context, *Oceanogr. Mar. Biol. Annu. Rev.*, 18, 423–481.

Hylleberg, J., and Gallucci, V. 1975. Selectivity in feeding by the deposit bivalve *Macoma nasuta, Mar. Biol.*, 32, 167–178.

Jørgensen, C.B. 1966. *Biology of Suspension Feeding*, Pergamon Press, Oxford, 357 pp.

Jørgensen, C.B. 1990. *Bivalve Filter Feeding: Hydrodynamics, Bioenergetics, Physiology, and Ecology*, Olsen and Olsen, Eds., Fredensborg, Denmark, 140 pp.

Jørgensen, C.B., Larsen, P.S., Møhlenberg, F., and Riisgard, H.U. 1986. The bivalve pump, *Mar. Ecol. Prog. Ser.*, 34, 69–77.

Jumars, P., Self, R.F.L., and Nowell, A.R.M. 1982. Mechanics of particle selection by tentaculate deposit-feeders, *J. Exp. Mar. Biol. Ecol.*, 64, 47–70.

Kauffman, E.G. and Sohl, N.F. 1974. Structure and evolution of Cretaceous Rudist frameworks, *Verh. Naturforsch. Ges.*, 84, 399–467.

Kenk, V.C. and Wilson, B.R. 1985. A new mussel (Bivalvia, Mytilidae) from hydrothermal vents in the Galapagos Rift zone, *Malacologia*, 26, 253–277.

Kinne, O. 1964. The effects of temperature and salinity on marine and brackish-water animals. II. Salinity and temperature-salinity combinations, *Oceanogr. Mar. Biol. Annu. Rev.*, 2, 281–339.

Kiørboe, T. and Møhlenberg, F. 1981. Particle selection in suspension-feeding bivalves, *Mar. Ecol. Prog. Ser.*, 5, 291–296.

Kiørboe, T. Møhlenberg, F., and Nøhr, O. 1981. Effect of suspended bottom material on growth and energetics in *Mytilus edulis*, *Mar. Biol.*, 61, 283–288.

Klumpp, D.W., Bayne, B.L., and Hawkins, A.J.S. 1992. Nutrition in the giant clam *Tridacna gigas* (L.). I. Contribution of filter feeding and photosynthates to respiration and growth, *J. Exp. Mar. Biol. Ecol.*, 155, 105–122.

Kreeger, D.A., Langdon, C.J., and Newell, R.I.E. 1988. Utilization of refractory cellulosic carbon derived from *Spartina alterniflora* by the ribbed mussel *Geukensia demissa*, *Mar. Ecol. Prog. Ser.*, 42, 171–179.

Kruger, F. 1960. Zur Frage der Grossenabhangigkeit des Sauerstoff verbrachs von *Mytilus edulis, Helgo. Wiss. Meeresunters.*, 7, 125–148.

Langdon, C.J. and Newell, R.I.E. 1990. Utilization of detritus and bacteria as food sources by two bivalve suspension-feeders, the oyster *Crassostrea virginica* and the mussel *Geukensia demissa*, *Mar. Ecol. Prog. Ser.*, 58, 299–310.

Le Pennec, M. and Prieur, D. 1984. Observations sur la nutrition d'un Mytilidae d'un site hydrothermal actif de la dorsale du Pacifique Oriental, *C. R. Acad. Sci. Paris Ser. III*, 298, 493–500.

Levinton, J.S. 1982. *Marine Ecology*, Prentice Hall, Englewood Cliffs, NJ, 526 pp.

Loomis, S.H., Carpenter, J.F., and Crowe, J.H. 1988. Identification of strombine and taurine as cryoprotectants in the intertidal bivalve *Mytilus edulis, Biochim. Biophys. Acta*, 943, 113–118.

Lopez, G.R. and Crenshaw, M.A. 1982. Radiolabelling of sedimentary organic matter with ^{14}C-formaldehyde: preliminary evaluation of a new technique for use in deposit-feeding studies, *Mar. Ecol. Prog. Ser.*, 8, 283–289.

Lopez, G.R. and Levinton, J.S. 1987. Ecology of deposit-feeding animals in marine sediments, *Q. Rev. Biol.*, 62, 235–260.

Lutz, R.A. and Rhoads, D.C. 1977. Anaerobiosis and a theory of growth line formation, *Science*, 198, 1222–1227.

Lutz, R.A. and Rhoads, D.C. 1980. Growth patterns within the molluscan shell, in Rhoads, D.C. and Lutz, R.A., Eds., *Skeletal Growth of Aquatic Organisms*, Plenum Press, New York, 203–254.

Mangum, C. and Van Winkle, W. 1973. Responses of aquatic invertebrates to declining oxygen conditions, *Am. Zool.*, 13, 529–541.

Martin, A.W. and Harrison, F.M. 1964. Excretion, in Wilbur, K.M. and Yonge, C.M., Eds., *Physiology of Mollusca*, Academic Press, New York, 353–386.

Møhlenberg, F. and Riisgard, H.U. 1978. Efficiency of particle retention in 13 species of suspension feeding bivalves, *Ophelia*, 17, 238–246.

Morse, D.C. 1944. Some observations on the food and feeding of oysters in Chesapeake Bay, *Proc. Natl. Shellfish. Assoc.*, 55, 17–19.

Morton, B. 1978. Feeding and digestion in shipworms, *Oceanogr. Mar. Biol. Annu. Rev.*, 16, 107–144.

Muscatine, L. 1967. Glycerol excretion by symbiotic algae from corals and *Tridacna* and its control by the host, *Science*, 156, 516–519.

Newell, R.C. 1965. The role of detritus in the nutrition of two marine deposit feeders, the prosobranch *Hydrobia ulvae* and the bivalve *Macoma balthica*, *Proc. Zool. Soc. Lond.*, 144, 25–45.

Newell, R.C. and Field, J.G. 1983. The contribution of bacteria and detritus to carbon and nitrogen flow in a benthic community, *Mar. Biol. Lett.*, 4, 23–36.

Newell, R.I.E. and Langdon, C.J. In press. The eastern oyster: mechanisms and physiology of larval and adult feeding, in Eble, A., Kennedy, V.S., and Newell, R.I.E., Eds., *The Eastern Oyster, Crassostrea virginica*, Maryland Sea Grant Publications, MD.

Palmer, R. and Williams, L.G. 1980. Effect of particle concentration on filtration efficiency of the Bay Scallop *Argopecten irradians* and the oyster *Crassostrea virginica*, *Ophelia*, 19, 163–174.

Pamatmat, M.M. 1979. Anaerobic heat production of bivalves (*Polymesoda caroliniana* and *Modiolus demissus*) in relation to temperature, body size, and duration of anoxia, *Mar. Biol.*, 53, 223–229.

Peters, R.H. 1983. *The Ecological Implications of Body Size*, Cambridge University Press, New York, 329 pp.

Peterson, B.J., Howarth, R.W., and Garritt, R.H. 1985. Multiple stable isotopes used to trace the flow of organic matter in estuarine food webs, *Science*, 227, 1361–1363.

Porter, J.W. 1980. Primary productivity in the sea, reef corals *in situ*, *Brookhaven Symp. Biol.*, 31, 403–410.

Price, T.J., Ghayer, G.W., LaCroix, M.W., and Roberts, T.G. 1970. The organic content of shells, tissues and pallial fluid of some estuarine molluscs and its ecological implication, *Annu. Rep. Beaufort Fish. Lab.*, 186–190.

Prieur, D. 1981. Experimental studies of trophic relationships between marine bacteria and bivalve molluscs, *Kieler Meeresforsch., Sonderh.*, 5, 376–383.

Putter, A. 1909. *Die Ernahrung der Wasswertiere and der Stoffhaushalt der Gewasser*, Fisher, Jena.

Rao, K.P. 1953. Shell weight as a function of intertidal height in a littoral population of pelecypods, *Experientia*, 9, 465–468.

Reid, R.G.B. 1980. Aspects of the biology of a gutless species of *Solemya* (Bivalvia: Protobranchia), *Can. J. Zool.*, 58, 386–393.

Reid, R.G.B. and Brand, D.G. 1986. Sulfide-oxidizing symbiosis in Lucinaceans: implications for bivalve evolution, *Veliger*, 29, 3–24.

Reid, R.G.B., Fankboner, P.V., and Brand, D.G. 1984. Studies on the physiology of the giant clam *Tridacna gigas*. I. Feeding and digestion, *Comp. Biochem. Physiol.*, 78A, 95–101.

Rhoads, D.C. and Pannella, G. 1970. The use of molluscan shell growth patterns in ecology and paleoecology, *Lethaia*, 3, 143–161.

Rice, M.A. and Stephens, G.C. 1987. Uptake and internal distribution of exogenously supplied amino acids in the Pacific oyster *Crassostrea gigas*, *Aquaculture*, 66, 19–31.

Richards, F.J. 1959. A flexible growth function for empirical use, *J. Exp. Bot.*, 10, 290–300.

Rubenstein, D.I. and Koehl, M.A.R. 1977. The mechanisms of filter feeding: some theoretical considerations, *Am. Nat.*, 111, 981–994.

Seed, R. 1976. Ecology, in Bayne, B.L., Ed., *Marine Mussels: Their Ecology and Physiology*, Cambridge University Press, New York, 13–65.

Seed, R. and Suchanek, T.H. 1992. Population and community ecology of *Mytilus*, in Gosling, E., Ed., *The Mussel Mytilus: Ecology, Physiology, Genetics and Culture*, Elsevier, Amsterdam, 87–169.

Seiderer, L.J., Davis, C.L., Robb, F.T., and Newell, R.C. 1984. Utilisation of bacteria as nitrogen resource by kelp-bed mussel *Choromytilus meridionalis*, *Mar. Ecol. Prog. Ser.*, 15, 109–116.

Sherr, E.B., Sherr, B.F., Fallon, R.D., and Newell, S.Y. 1990. Small aloricate ciliates as a major component of the marine heterotrophic nanoplankton, *Limnol. Oceanogr.*, 37, 177–.

Shick, J.M., Zwaan, A. de and Bont, A.M. de, 1983. Anoxic metabolic rate in the mussel *Mytilus edulis* estimated by simultaneous direct calorimetry and biochemical analysis, *Physiol. Zool.*, 56, 56–63.

Shumway, S.E., Cucci, R.C., Newell, R.C., and Yentsch, C.M. 1985. Particle selection, ingestion, and absorption in filter-feeding bivalves, *J. Exp. Mar. Biol. Ecol.*, 91, 77–92.

Sleigh, M.A. and Aiello, E. 1972. The movement of water by cilia, *Acta Protozool.*, 11, 265–277.

Stephens, G.C. 1982. Recent progress in the study of "Die Ernahrung der Wassertiere und der Stoffhaushlt der Gewasser," *Am. Zool.*, 22, 611–619.

Stuart, V., Field, J.G., and Newell, R.C. 1982. Evidence for absorption of kelp detritus by mussel *Aulacomya ater* using a new ^{51}Cr-labeled microsphere technique, *Mar. Ecol. Prog. Ser.*, 9, 263–271.

Taghon, G.L. and Jumars, P.A. 1984. Variable ingestion rate and its role in optimal foraging behavior of marine deposit feeders, *Ecology*, 65, 549–558.

Taylor, C.C. 1959. Temperature and growth — the Pacific razor clam, *J. Cons.*, 24, 93–101.

Taylor, C.C. 1960. Temperature, growth, and mortality — the Pacific cockle, *J. Cons.*, 25, 117–124.

Thompson, D.W. 1961. *On Growth and Form*, Cambridge University Press, New York.

Thompson, R.J. 1979. Fecundity and reproductive effort of the blue mussel (*Mytilus edulis*), the sea urchin (*Strongylocentrotus droebachiensis*) and the snow crab (*Chionectes opal*) from populations in Nova Scotia and Newfoundland, *J. Fish. Res. Board Can.*, 36, 955–964.

Trench, R.K., Wethey, D.S., and Porter, J.W. 1981. Observations on the symbiosis with zooxanthellae among the Tridacnidae, *Biol. Bull.*, 161, 180–198.

Ursin, E. 1963. On the incorporation of temperature in the von Bertalanffy growth equation, *Medd. Dan. Fisk. Havunders. N.S.*, 4, 1–16.

Vahl, O. 1972. Efficiency of particle retention in *Mytilus edulis*, *Ophelia*, 10, 17–25.

Walford, L.A. 1946. A new graphic method of describing the growth of animals, *Biol. Bull.*, 90, 141–147.

Walne, P.R. 1972. The influence of current speed, body size and water temperature on the filtration rate of five species of bivalves, *J. Mar. Biol. Assoc. U.K.*, 52, 345–274.

Ward, J.E., MacDonald, B.A., Thompson, R.J., and Beninger, P.G. 1993. Mechanisms of suspension feeding in bivalves: resolution of current controversies by means of endoscopy, *Limnol. Oceanogr.*, 38, 265–272.

Ward, J.E., Newell, R.I.E., Thompson, R.J., and MacDonald, B.A. 1994. *In vivo* studies of suspension-feeding processes in the Easter oyster, *Crassostrea virginica* (Gmelin), *Biol. Bull.*, 186, 221–240.

Warren, C.E. and Davis, G.E. 1967. Laboratory studies on the feeding bioenergetics and growth of fish, in Gerking, S.D., Ed., *The Biological Basis of Freshwater Fish Production*, Blackwell Scientific, Oxford, 175–214.

Waterbury, J.B., Turner, R.D., and Calloway, C.B. 1983. A cellulolytic nitrogen-fixing bacterium cultured from the gland of Deshayes in shipworms, *Science*, 221, 1401–1403.

Widdows, J. 1976. Physiological adaptation of *Mytilus edulis* to cyclic temperatures, *J. Comp. Physiol.*, 105, 115–128.

Widdows, J. and Bayne, B.L. 1971. Temperature acclimation of *Mytilus edulis* with reference to its energy budget, *J. Mar. Biol. Assoc. U.K.*, 51, 827–843.

Wilbur, K.M. 1964. Shell formation and regeneration, in Wilbur, K.M. and Yonge, C.M., Eds., *Physiology of Mollusca*, Academic Press, New York, 243–282.

Wilbur, K.M. and Saleuddin, S.M. 1983. Shell formation, in Saleuddin, S.M. and Wilbur, K.M., Eds., *The Mollusca*, Vol. 4, *Physiology*, Part 1, Academic Press, New York, 236–287.

Wilkerson, F.P. and Trench, R.K. 1986. Uptake of dissolved inorganic nitrogen by the symbiotic clam *Tridacna gigas* and the coral *Acropora*, *Mar. Biol.*, 93, 237–246.

Williams, L.G. 1982. Mathematical analysis of the effects of particle retention efficiency on determination of filtration rate, *Mar. Biol.*, 66, 171–177.

Willows, R.I. 1992. Optimal digestive investments: a model for filter feeders experiencing variable diets, *Limnol. Oceanogr.*, 37, 829–847.

Winter, J.E. 1969. Über den Einfluss der Nahrungskonzentration und anderer Faktoren auf Filtrierleistung und Muschlen *Arctica islandica* und *Modiolus modiolus, Mar. Biol.*, 4, 87–135.

Winter, J.E. 1970. Filter feeding and food utilization in *Arctica islandica* and *Modiolus modiolus* at different food concentrations, in Steele, J.H., Ed., *Marine Food Chains*, Oliver and Boyd, Edinburgh, 196–206.

Winter, J.E. 1973. The filtration rate of *Mytilus edulis* and its dependence on algal concentration measured by a continuous automatic recording apparatus, *Mar. Biol.*, 22, 317–328.

Winter, J.E. 1978. A review on the knowledge of suspension-feeding in lamellibranchiate bivalves, with special reference to artificial aquaculture systems, *Aquaculture*, 13, 1–33.

Wright, R.T., Coffin, R.B., Ersing, C.P., and Pearson, D. 1982. Field and laboratory measurements of bivalve filtration of natural marine bacterioplankton, *Limnol. Oceanogr.*, 27, 91–98.

Wright, S.H. 1982. A nutritional role for amino acid transport in filter-feeding marine invertebrates, *Am. Zool.*, 22, 621–634.

Wright, S.H. 1985. Mechanism and physiological role of epidermal amino acid transport in marine bivalves, in Gilles, R and Gilles-Bailien, M., Eds., *Comparative Physiology and Biochemistry: Current Topics and Trends*; Vol. B, *Transport, Iono- and Osmoregulation*, Springer-Verlag, Berlin, 292–302.

Yamamoto, N. and Lopez, G. 1985. Bacterial abundance in relation to surface area and organic content of marine sediments, *J. Exp. Mar. Biol. Ecol.*, 90, 209–220.

Yonge, C.M. 1936. Mode of life, feeding, digestion and symbiosis with zooxanthellae in the Tridacnidae, *Sci. Rep. Gr. Barrier Reef Exped. Br. Mus.*, 1, 283–321.

Yonge, C.M. 1947. The pallial organs in the aspidobranch Gastropoda and their evolution throughout the Mollusca, *Phil. Trans. Roy. Soc. Lond. Ser. B*, 232, 443–518.

Yonge, C.M. 1975. Giant clams, *Sci. Am.*, 23, 96–105.

Zobell, C.E. and Landon, W.A. 1937. Bacterial nutrition of the California mussel, *Proc. Soc. Exp. Biol. Med.*, 36, 607–609.

Zobell, C.E. and Feltham, C.B. 1938. Bacteria as food for certain marine invertebrates, *J. Mar. Res.*, 1, 312–327.

Zwaan, A. de 1977. Anaerobic energy metabolism in bivalve molluscs, *Annu. Rev. Oceanogr. Mar. Biol.*, 15, 103–187.

Zwaan, A. de and Mathieu, M. 1992. Cellular biochemistry and endocrinology, in Gosling, E., Ed., *The Mussel Mytilus: Ecology, Physiology, Genetics and Culture*, Elsevier, Amsterdam, 223–307.

4 Population Processes

INTRODUCTION

A group of organisms of the same species living in the same area is termed a population. Each population has a number of characteristics that can be categorized as biological or group attributes. Biological attributes are characteristics the population shares with its component organisms. Life history or cycle is most often given as a biological attribute because a population grows, differentiates, and maintains itself just like individual organisms. Group attributes are common only to the population and include birth rate (natality), death rate (mortality), and age structure. These group attributes are unique to the population and only have meaning to a population of organisms (Odum 1983). These population or group attributes reflect the ecology of a species and can provide information on the spatial and temporal aspects of the environment (Cerrato 1980).

In this chapter, the demography of bivalve populations will be presented by examining life tables of specific bivalves and by exploring a number of population processes, i.e., growth, recruitment, and mortality, for a variety of species. As all of these processes influence the production of bivalves, understanding them is vital to determining the role of these animals in ecosystems.

LIFE CYCLE

The life cycle of bivalve molluscs begins with fertilization of eggs, usually in the water surrounding the adults. Gametes are typically shed within the shell cavity and swept out on the exhalant currents generated by pumping. In contrast, some bivalves brood their eggs, with fertilization occurring in the mantle cavity. Here again the bivalve water pump is important in bringing in the sperm for fertilization. In shipworms, brooding occurs in the suprabranchial cavity, and in the oyster, *Ostrea*, within the gills (Barnes 1974).

The eggs of bivalves develop into free-swimming trochophore and then veliger larvae. In some marine bivalves, the veliger stage may last as long as several months. In others, it is only a few days. In brooders, the veliger larvae are released after a specific period (8 days in *Ostrea edulis*). The larval stage is planktonic and subject to very high mortality. At metamorphosis, the larvae settle out of the plankton and the bivalve spends its remaining life as a member of the benthos. Consequently, the life cycle of a marine bivalve can be separated into two very distinct stages, the planktonic presettlement–larval period and the post settlement juvenile–adult period. The presettlement period is important for dispersal of the species, while the juvenile–adult period is important for reproduction. The two stages are spatially and temporally separated for most of their existence, and that reduces the chances of adult induced mortality on the larvae through adult filter feeding.

The population dynamic of the presettlement stage is poorly studied, probably because of its short duration and the greater logistical complications of sampling plankton. Thus, most studies of bivalve population dynamics focus on the post-settlement stage and particularly on shallow water, commercially valuable species.

STATISTICAL MEASURES OF POPULATIONS

Postsettlement bivalves are excellent subjects for population studies because their shells provide a long lasting record of the organism's life and, through statistical compilation, the dynamics (demography) of a given population. The shell is an easily observed and identified characteristic of a bivalve species and collections of living animals can yield accurate estimates of population density (Cerrato 1980). In some cases, shells from ancient deposits can also be used to determine the densities and other properties of past populations of bivalves (Rhoads et al. 1982; Powell et al. 1984). Through shell growth rings, the growth rate and age of individual bivalves can be assessed. This information can be compiled to determine age structure for a specific population. In some cases, the shell records tell-tale clues as to the causes of mortality. Not all population parameters can be determined from the shell. Fecundity requires the analysis of living bivalves, while incidences of disease and parasitism may require direct observation of living tissues.

DENSITY

Often the simplest population attributes to determine are density and abundance. Generally, density refers to the population size at a given time, i.e., number of individuals per unit area, while abundance implies a dynamic (changing) aspect to the population, i.e., number per unit area per unit time. Because the individuals in a population vary physiologically and genetically and respond differently to the environment, it is usually more meaningful to focus on the population at a level of resolution greater than simple density.

Density observations on a population are easily arranged into nonoverlapping groups or classes that divide the life span or size range of the organism. The class frequencies for the data are found by summing the number of individuals in each class. The class frequency data may appear as a table or more often as a histogram with classes (size or age) plotted versus frequency (Figure 22). In bivalves, age is often used in these plots, as growth or age can be clearly determined from rings on the shell. If small size class intervals are used in constructing the histogram, variations in growth within a year or recruitment class can be resolved (Cerrato 1980).

Most temperate and boreal shallow water bivalves are seasonally programmed with distinct spawning periods. With frequent sampling, these spawning periods will be reflected in the size frequency histograms as a pulse of new individuals at the smallest size class. Normally, the smaller size or year classes make up the numerical majority of the population while the older and larger classes are only a small portion

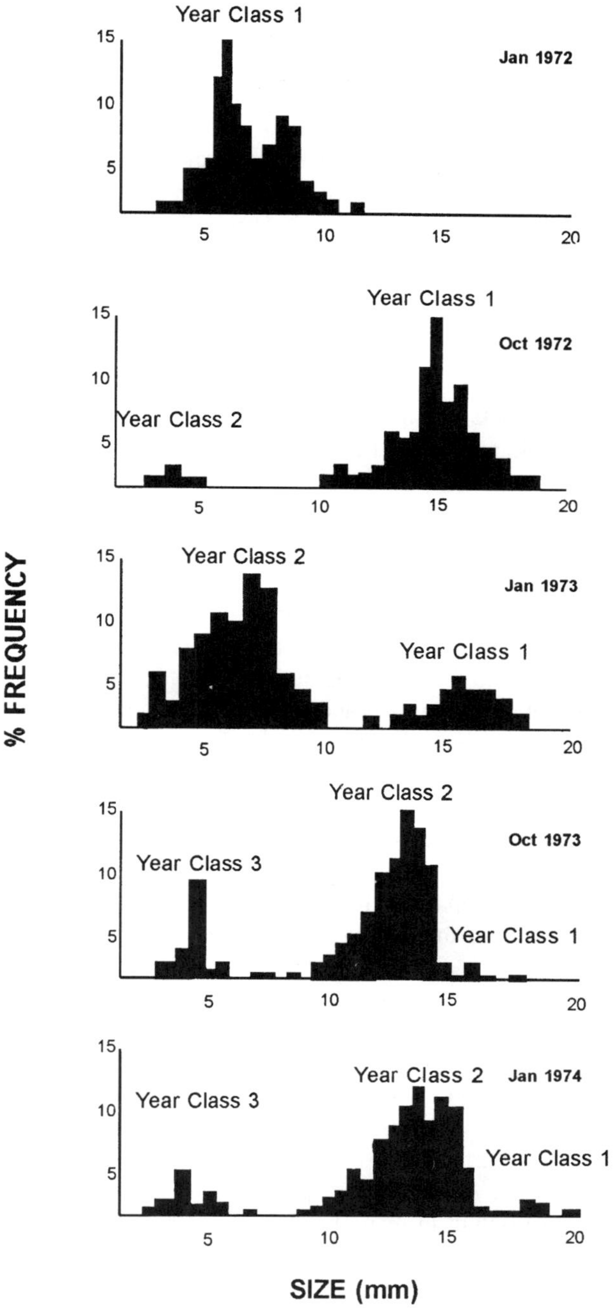

FIGURE 22. Size-frequency distributions of the cockle *Cerastoderma* in Ireland. (Adapted from Seed, R. and Brown, R.A. 1975.)

of the total population size. If additional samples are taken through time, these samples will show recruitment pulses migrating across successive size frequency plots. This movement across the frequency plots is caused by the growth of individuals within the population and offers a way of computing modal or average growth for a size class (see Chapter 3). Mortality will also cause the pulse to decrease in height or shrink and it too can be estimated from the size frequency plot (see Figure 22).

For long-lived species, multiple pulses are often discernible in the size frequency histogram, and each pulse represents an age class or year class in the population. It is also common for an age class or pulse to have a roughly normal distribution and to overlap adjacent age classes. There are a number of graphical and statistical methods (Walford 1946; Cassie 1954; Sheldon 1965; Cerrato 1980; Basson et al. 1988) that can be used to separate the overlapping age classes. Thus, the size frequency histogram contains valuable information about the population in question, over and above simple measures of density.

REPRODUCTION

The process of reproduction is the generation of new individuals that have the potential to become members of the population. Reproduction is integrated into the bivalve's life cycle and is one of the few instances where the planktonic gamete and larval stages are linked to the postsettlement juvenile and adult stages. There are a number of terms that are often used and confused when discussing reproduction in bivalves at the population level, and they need to be specifically defined (after Krebs 1978).

Fecundity is the potential capability of an organism to produce reproductive units.

Fertility is the actual capability for an organism to produce offspring.

Natality or *birth rate* is the rate of production of new individuals by a population.

Reproductive output is the amount of gametes released into the environment.

Settlement is the process by which individual bivalve larvae come into contact with and attach to the bottom (usually permanently).

Recruitment is the successful colonization of the bottom by bivalves and implies the passage of time with survival and postsettlement mortality (Seed and Suchanek 1992).

The reproductive output in many bivalves can represent a major proportion of total population production and can be a significant energy input to the water column. Most bivalves are iteroparous (multiple breeding in a lifetime) and their reproductive output varies according to size or age, with smaller individuals putting proportionally more effort into somatic growth and older individuals putting more effort into reproduction (Bayne 1976). Therefore, younger individuals as a group may contribute as many or more offspring because their numbers are usually much greater than those of larger individuals in the population. This change in reproductive output with age appears to be the rule not only for energy and carbon, but also for nitrogen (Rodhouse et al. 1984). Reproductive output or gamete production in most bivalves seems to follow the planktotrophic strategy (Vance 1973; Bayne 1976) that is characterized by high fecundity and high metabolic cost.

A cursory survey of estimated bivalve reproductive effort as a proportion of total production is given in Table 5. Clearly, reproductive effort is highly variable both within and among species. While most low values can be related to young populations, those values for *Crassostrea* and *Geukensia* are for mature populations. However, these populations are in the mid- to upper intertidal zone in the subtropical U.S. marsh–estuarine ecosystems where stress due to exposure may detract energy from reproductive effort. Changes in year-to-year environmental conditions may also explain variations in population reproductive effort in *Tellina, Choromytilus,* and *Perna*.

TABLE 5
Bivalve Reproductive Effort in Relation to Total Production (%)

Species	Reproductive Effort (%)	Ref.
Aulacomya ater	63	Griffiths and Griffiths 1987
Chione cancellata	50	Moore and Lopez 1969
Choromytilus meridionalis	73	Griffiths 1981
Crassostrea gigas	98	Bernard 1974
Crassostrea virginica	16	Dame 1976
Dosinia elegans	47	Moore and Lopez 1970
Geukensia demissa	17	Kuenzler 1961
Lyrodus pedicellatus	47	Gallager et al. 1981
Mercenaria mercenaria	46	Hibbert 1977
Ostrea edulis	47	Rodhouse 1978
Patinopectin yessoensis	3–38	Fuji and Hashizuma 1974
Perna perna	16	Berry 1978
Scrobicularia plana	24–52	Hughes 1970
Tellina tenuis	9–62	Trevallion 1971

POPULATION GROWTH

The size of a population in terms of numbers is a function of the inputs and outputs to that population.

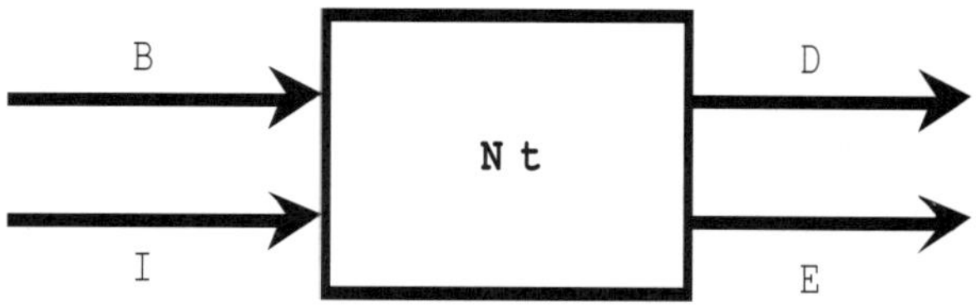

The major inputs or additions to the population are the births of new individuals (B) and immigration (I), while the major losses or outputs are death (D) and emigration (E). The size or density of the population is $N_{t.}$ The change in population size (ΔN) per unit change in time (Δt) can be represented by:

$$\frac{\Delta N}{\Delta t} = B + I - (D + E)$$

From an ecosystem perspective, it is probably more meaningful to represent these parameters as biomass in units of energy or grams of material, i.e., carbon, nitrogen, or phosphorus (Valiela 1984).

When a population exists in a favorable environment with no limitations due to space, food, predation, or environment, the specific growth rate (growth rate per individual) of the population is constant and maximal for the specific conditions. This growth rate is specific for a particular population age structure, as found in a size class histogram, and is a size index of population growth (Odum 1983). This growth is represented by r, a coefficient in the exponential growth model:

$$\frac{dN}{dt} = rN$$

and where

$$r = dN(Ndt)$$

or when integrated

$$N_t = N_o e^{rt}$$

Where N_t is the population size at some later time t, N_o is the population size at the initial time, e is the exponential constant and r is the growth coefficient. Taking the natural logarithm of the preceding equation

$$\ln N_t = \ln N_o + rt$$

and

$$r = \frac{\ln N_t - \ln N_o}{t}$$

The index r is, then, the difference between the instantaneous specific birth (natality) rate b and the instantaneous death (mortality) rate d. When emigration and immigration are small or nonexistent, as is the case with most bivalves, then they can be ignored and

$$r = b - d$$

This form of r is called the intrinsic rate of natural increase or r_{max} and is dependent on a stable and stationary population age structure.

Brousseau (1978) has calculated r for the clams *Mya arenaria* from the relationship:

$$1 = \sum_{0}^{n} \exp^{-rx} l_x m_x$$

Given the values for l_x, m_x and r (see Table 6 for definitions), Brousseau (1978) used Fisher's (1958) equation

$$\frac{v_x}{v_o} = \frac{e^{rx}}{l_x} \int_{y=x}^{\infty} e^{-ry} l_y m_y d_y$$

to compute the reproductive values v_x/v_o (Table 6). This integral is the sum of the number of offspring produced by a female for a given age (x) as this organism passes through all the remaining age classes (y). These values indicate at what age a larva will establish a successful colony and are shown in Figure 23 for *Mya*. As r increases, the importance of older age categories increases with regard to reproductive value and is largely due to increasing reproductive rate in older clams. Thus, r can be used to generate additional information from the data compiled in a life or fecundity table.

TABLE 6
Life and Fecundity Table Definitions and Calculations

Property	Definition	Observation or Calculation
x	Age interval	Age or size class data from histogram of live or dead shells
n_x	Number of individuals at start of age interval	
l_x	Proportion of individuals surviving to start age interval	Usually adjusted to a cohort of 1000 for age interval 1
d_x	Number of individuals dying during an age interval	$l_x + l_{x+1}$
q_x	Rate of mortality during an age interval	d_x/l_x
L_x	Average number of individuals alive during an age interval	$[l_x - l_{x+1}]/2$
T_x	Number of individual time units left in a population at the start of an age interval	$\sum_{x}^{\infty} L_x$
e_x	Average life expectancy for an age interval	T_x/l_x
m_x	Age-specific fecundity (number of larvae spawned)	Estimated from oocyte production or plankton
V_x	Expected number of offspring	$l_x m_x$

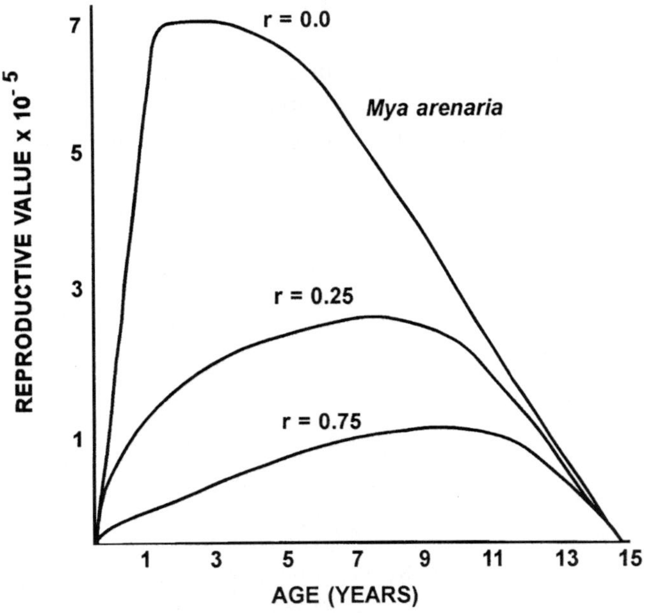

FIGURE 23. The relationship between reproductive value and age for r values for the clam *Mya arenaria*. (Adapted from Brousseau, D.J. 1978.)

BIVALVES AS OPEN OR METAPOPULATIONS

As the spatial scale for considering a population increases, the population dynamics are less affected by movements (immigration and emigration) and become increasingly dominated by the interactions of birth and death rates discussed above (McGrorty and Goss-Custard 1991). With organisms that have a great potential for dispersal, e.g., marine bivalve larvae, enormous geographical areas must be considered in order to achieve the birth rate and death rate control of population dynamics. This requirement is usually impractical. Instead of this closed or entire population approach, the combined population properties of many local groups (patches, subpopulations, etc.) or metapopulations can be utilized and depends not only on the interaction between groups (Roughgarden et al. 1985) but also within group processes of density-dependent mortality, self-thinning, density-limited growth, and density-dependent emigration (McGrorty and Goss-Custard 1991). With this approach, the distribution, size, and density of the metapopulation may be regulated or limited by the supply of larvae from the plankton rather than by local fecundity (McGrorty et al. 1990). In a field study of the population dynamics of *Mytilus edulis* in the Exe estuary, England, McGrorty et al. (1990) and McGrorty and Goss-Custard (1991) found that annual variations in spat fall over the entire estuary were related closely to the densities of adults on the different beds. They further found that the main environmental gradient along which the population dynamics of mussels in this estuary survived was exposure time. Mortality regulation during the first year resulted in a pronounced stability of the total population.

LIFE AND FECUNDITY TABLES

The age-specific statistical characteristics of mortality, survivorship, and life expectancy are generally represented in columnar form in a life table. Because the size of a population may change as a function of births and deaths of individuals, the life table provides an organized method for determining these properties.

The life table is made up of rows and columns of both observed and calculated information on a population. The rows usually represent the age or year classes, while the columns denote various population properties. Definitions and calculations for the columns are given in Table 6.

While the number of individuals alive at a particular time (n_x) is directly observed, mortality can be obtained from age of death observations by computing the decline in the numbers of individuals in an age class through time and by directly monitoring the death rate of marked individuals (Cerrato 1980). It is normal to analyze mortality by focusing on one property, l_x, the survivorship. Usually the log of survivorship is plotted as a function of age (x). Although there are several standard survivorship curve forms, Kurten (1964) has pointed out that when the complete life cycle of a species is considered, the shape of the survivorship curve is sigmoidal, with high mortality in both young and old age classes (Figure 24).

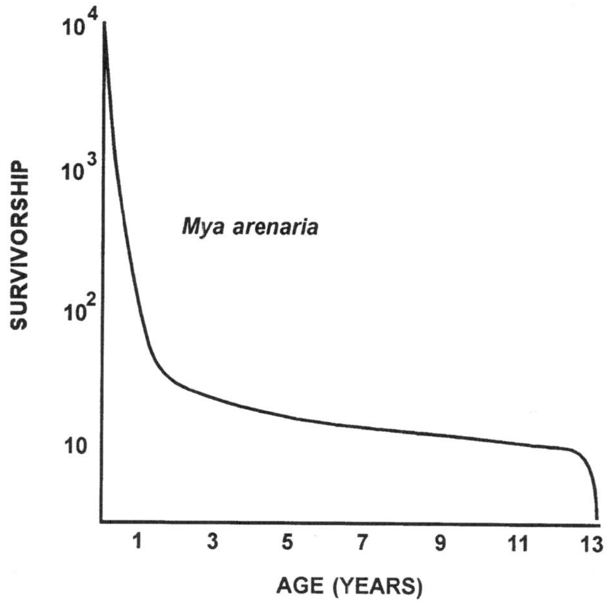

FIGURE 24. A survivorship curve for the clam *Mya arenaria* at Gloucester, MA. (Adapted from Brousseau, D.J. 1978.)

A classical example of a bivalve life table is given in Table 7. In the Pismo clam, *Tivela stultorum*, the numbers of observed living clams in a given age interval are adjusted to an initial cohort of 1000 juvenile clams. Of this initial cohort, 550 clams die during the first age interval (d_x) and can be expressed as a mortality rate (q_x) of

TABLE 7
Life Table for *Tivela stultorum,* the Pismo Clam

Age X (years)	l_x	d_x	q_x	L_x	T_x	e_x (years)
0–1	1000	550	550	723	1569	1.57
1–2	450	202	449	349	844	1.87
2–3	248	72	290	212	495	1.99
3–4	176	60	341	146	283	1.60
4–5	116	60	517	86	137	1.18
5–6	56	38	678	37	51	0.90
6–7	18	13	722	11.5	14	0.75
7–8	5	5	1000	2.5	2.5	0.50

After Cerrato, R.M. 1980.

550/1000 or 0.550. The average number of clams alive during the age interval (L_x) is 723 or (1000 + 450)/2. Summing the L_x column from the bottom yields the T_x value, and dividing this by l_x gives the life expectancy in years (e_x) for each age interval. The data show that young (0 to 2) and older (4 to 8) age interval clams die more rapidly than middle-aged clams. Life expectancy is highest for 1- 3-year-old clams. These observations are typical for invertebrate populations that do not brood or protect their young and experience high mortality after settlement.

An excellent example of a life table with fertility computations for a marine bivalve, *Mya arenaria*, is given by Brousseau (1978) and is described in Table 8. The data in this table are presented differently than those in Table 7 in that the proportion of clams surviving (l_x) is used rather than observed numbers and the first three year classes were subdivided. This life table (Table 8) shows that *Mya* reaches maximum life expectancy at 30 mm length or about an age of 1 year. Mortality as described by survivorship is highest for larvae and very old size classes (Figure 24). This survivorship curve form is sigmoidal as suggested by Kurten (1964), and supports Murphy's (1968) theory that long life and iteroparity are favored under conditions where spawning or settlement success are highly variable. Using histological analysis, Brousseau (1978) also was able to compute age-specific fecundity. This property increased with size and age from 40 mm and 1.7 years with a fecundity of 7399 and to a maximum age and size with a fecundity of 72,125 or an order of magnitude higher than the younger clams. Using her estimates of l_x and m_x, Brousseau (1978) was able to compute a very high intrinsic rate of natural increase (r_{max}) of 4.74 for this clam population. This enormous potential for increase is completely offset by high larval mortality, i.e., in the order of 0.001 to 0.0001% of larvae spawned eventually settled (Brousseau 1978).

As we have seen in the size frequency histograms, new recruits to the bivalve population show up as pulses of individuals in the smallest size or year classes. Powell et al. (1984) have shown that if the sample frequency is too long the

TABLE 8
***Mya arenaria* Life and Fecundity Table**

Size class (mm)	Age x (years)	l_x	L_x	e_x (years)	q_x	m_x
2.0–4.9	0.143	0.557		0.769	0.885	—
5.0–9.9	0.286	0.662	0.610	1.820	0.675	—
10.0–14.9	0.428	0.800	0.731	3.562	0.400	—
15.0–19.9	0.571	0.800	0.800	4.602	0.400	—
20.0–24.9	0.714	0.834	0.817	6.336	0.333	—
25.0–29.9	0.857	0.900	0.867	8.248	0.199	—
30.0–34.9	1.000	0.968	0.934	9.178	0.062	—
35.0–39.9	1.333	0.978	0.973	8.758	0.045	—
40.0–44.9	1.667	0.963	0.971	8.143	0.075	7,379
45.0–49.9	2.000	0.966	0.965	7.758	0.066	13,589
50.0–54.9	2.333	0.973	0.970	7.268	0.055	—
55.0–59.9	2.666	0.962	0.968	6.665	0.078	18,835
60.0–64.9	3.000	0.952	0.957	6.186	0.098	33,605
65.0–69.9	4.000	0.949	0.951	5.794	0.101	36,624
70.0–74.9	5.000	0.969	0.959	5.386	0.059	46,034
75.0–79.9	6.000	0.984	0.977	4.690	0.031	56,986
80.0–84.9	7.000+	0.911	0.948	3.830	0.176	72,125

After Brousseau, D.J. 1978.

abundance of newly settled recruits will not be estimated accurately. This is because most juvenile bivalves recruited to the benthic population may settle, live, and die between sampling periods and thus, not be included in the size or age frequency histogram. These authors found that by sampling the shells of bivalves in the sediments adjacent to the live population, they could construct the death assemblage of the population. Input into the short-term death assemblage occurs in pulses and these pulses gradually decay as a result of taphonomic processes that destroy the individual shells. If larval settlement produces pulses of juveniles, if pulse decay rates are low, and if the half-lives of the decaying pulses are known, then the size of the original settlement can be determined from the death assemblage. From this analysis, Powell et al. (1984) determined that larval settlement is consistently underestimated by as much as 90% from living population data on bivalves.

MORTALITY

As noted earlier, the height of the age size class histograms in postsettlement bivalves decreases in size as mortality reduces population size. Mortality is also explicit in the plot of survivorship vs. time. Mortality in bivalves is caused by abiotic factors, predation, disease, and competition and human harvest for food or other products.

ABIOTIC MORTALITY

As discussed in Chapter 2, physical factors limit the existence of a bivalve to a given environment or niche. For bivalves, the most important of these abiotic mortality agents are temperature, aerial exposure, salinity, oxygen concentration, siltation, and waves. Depending on the latitude and coastal morphology, many of these factors may vary seasonally and, thus, result in seasonal mortality effects. In addition, many physical factors may act in concert with other factors to cause mortality. At the extreme, catastrophic environmental events, e.g., storms or widespread anoxia, may lead to mass mortality in bivalve populations.

Temperature can be a major cause of mortality in the intertidal zone. Summer extremes of high temperature during daytime low tide exposure can lead to extensive mortalities, particularly in subtropical and tropical populations of oysters (Dame 1976) and mussels (Suchanek 1978; Tsuchiya 1983), while extreme cold and icing conditions can do likewise in northern temperate and boreal bivalve populations (Suchanek 1986). Strong waves associated with storms can dislodge intertidal bivalves, resulting in mortality. Such conditions may also bring large floating debris, i.e., logs, that batter, crush, and dislodge bivalves in rocky shallow and intertidal environments (Dayton 1971; Lubchenco and Menge 1978; Suchanek 1986). In unconsolidated sedimentary environments, shifting of particles, burial, erosion, and direct movement of bivalves by currents and waves can also cause patchiness.

Anaerobic conditions are common to shallow bays and estuaries with poor circulation. Although some bivalves, e.g., oysters, can close their valves and resist anaerobic conditions for weeks, others (e.g., mussels) may suffer immediate mass mortality. For example, in the Dutch Oosterschelde estuary, large populations of mussels are cultivated. If the storm surge barrier separating the estuary from the North Sea is closed, water circulation is greatly reduced and the large population of mussels may rapidly drive the system to an anaerobic state and result in mass mortalities of mussels.

PREDATION

Often the most important cause of mortality in bivalve populations is predation. Predation can be defined as (a) the process that occurs when one organism kills another for food, or (b) when individuals of one species eat living individuals of another, or (c) a process by which one population benefits at the expense of another, or (d) any ecological process in which energy and matter flow from one species to another (Taylor 1984). For our purposes, predation is a direct interaction between different species populations.

In bivalve-dominated systems, the most common predators are gastropods, starfish, crabs, fish, birds, and mammals. Rather than be controlled by a specific predator, most sessile and slow-moving bivalves are usually preyed upon by suites or gilds of predators (Seed 1993). Many bivalve predators are only effective on certain size classes of prey, usually the smaller sizes, and at specific seasons of the annual cycle. Methods of predation range from defeating the defenses of the bivalve shell by small invertebrates and some birds, to crushing or breaking the shells by crabs, fish, and mammals, to the swallowing of the whole bivalve by birds. Most of our information on the predation of bivalves comes from intertidal and shallow water environments and is usually focused on economically important species. Those interested in a detailed review of the influence of invertebrate predators on bivalves should see Seed (1993) and of birds, Meire (1993).

Because bivalve larvae live in a different habitat, they suffer predation from an entirely different group of predators than do the benthic adult stages. The most commonly reported bivalve larval predators are ctenophores and jellyfish. In the bays along the Atlantic coast of North America, for example, the ctenophore *Mnemiopsis leidyi* has been deduced to be a major planktonic predator of oyster larvae as the ctenophore's density is inversely related to that of the oyster larvae (Kremer 1979). Curiously, this ctenophore is also preyed upon by the sea nettle, *Chrysaora quinquecirrha*, which is also a major predator of oyster larvae in Chesapeake Bay (Cargo and Shultz 1967).

In addition to these truly planktonic predators, benthic adult bivalve filter feeders may also ingest bivalve larvae (Korringa 1949). Although the idea that there is a significant interaction between established adult bivalve filter feeders and their settling larvae has been strongly questioned by Black and Peterson (1988), field manipulation studies with dense populations of the cockle, *Cerastoderma edule*, suggest that adult–larval interactions play an important role in this species population dynamics (Andre and Rosenberg 1991; Andre et al. 1993). Studies by Möller (1986) suggested that plots with high densities of *Mya arenaria* had fewer settled juveniles of *M. arenaria* and *C. edule* than control plots with low densities of *M. arenaria*. Andre and Rosenberg (1991) concluded that the settlement of bivalve larvae may be reduced by up to 40% in the presence of adult *C. edule* and *M. arenaria*. In a follow-up study using an experimental flume, Andre et al. (1993) determined that adult *C. edule* reduced the settlement of *C. edule* larvae by 33% in an area immediately surrounding individual adults. They also found that over dense beds of these cockles, as many as 75% of the observed larvae were inhaled by adults. These authors concluded that for *C. edule* beds the reduction in larval settlement was small to moderate at the scale of the individual adult, but that inhalation of settling larvae by the populations of resident filter feeding bivalves may cause a significant decrease in settlement on a larger scale (10 to 1000 m). Thus, when spatial and temporal scaling are taken into account, it appears that dense populations of adult bivalve filter feeders can be and are a major cause of mortality for settling larvae.

Compared to the predation of planktonic larvae, the predation of postsettlement bivalves is better studied because of the accessibility of the prey. The predators range in size from small flatworms to mammals. Rather than conduct a comprehensive review of bivalve predators (see Seed 1993 and Meire 1993), I shall provide a synopsis of the major players and give examples where appropriate.

The flatworm, *Stylochus ellipticus*, is an important predator of newly settled oysters on the Atlantic coast of North America (Loosanoff 1956 and Webster and Medford 1959). *Stylochus* appears to enter the oyster spat through the slightly opened valves where it then eats the soft body tissues. Although this flatworm seems to prefer barnacles, it will switch to oysters when they are abundant. Similar findings have been reported for *Stylochus mediterraneus* in mussels (Galleni et al. 1977).

Gastropod molluscs are common predators of bivalves. The majority of gastropod predators drill holes in the bivalve shell by a combination of mechanical (radula) and chemical means. These predators are often opportunistic, switching between prey species of bivalve or other phyla, e.g., barnacles, depending on prey availability and ease of successful attack. These predators also seem to favor younger or thinner shelled bivalves over older or thicker-shelled prey (Seed 1993). While most gastropod predation studies focus on temperate economically important bivalve species, i.e., clams, mussels and oysters (Galtsoff 1964; Seed 1992, 1993), studies of the cultivated tropical clam, *Tridacna maxima*, in the Cook Islands of the Pacific also reflect high gastropod predation on the smaller size classes (<10 cm) when the shell of *T. maxima* is thinner (Sims and Howard 1988). Thus, with increasing size and age, the thickening bivalve shell reduces the effectiveness of the gastropod predator. In essence, a size refuge is formed in the bivalve population for the very size classes that contribute the most to population production and reproduction.

Most crustacean predators are highly mobile and capable of searching for prey over large distances, even during a single tidal cycle (Seed 1993). They can feed on bivalves in both epibenthic and infaunal habitats as well as change food sources depending on availability. These decapods, mainly crabs and lobsters, use their specialized claws (chelae) and/or their mandibles to break the bivalve shell. Most decapod predators use more than one method to penetrate the bivalve shell. Lau (1987) and Seed (1993) have summarized these techniques in Table 9.

TABLE 9
Decapod Methods for Penetrating a Bivalve Shell

Method	Handling Time	Prey Size	Structural Investment
Swallowing whole	Short	Small	Low
Chipping–biting	Moderate	Medium	Low
Wedging	Short–moderate	Medium	Low
Prying	Moderate	Medium	Low
Crushing	Short	Medium–large	High
Chipping–peeling	Long	Medium–large	Moderate
Boring	Long	Medium	Moderate
Tubercular peeling	Long	Medium–large	Moderate
Shearing	Short	Medium	Moderate

After Lau (1987) and Seed (1993).

An illustrative example of decapod predation on bivalves was reported by Rhoads et al. (1982) for mussels inhabiting the deep sea hydrothermal vents along the Galapagos Rift in the Pacific Ocean. These investigators deduced that marks and damage to mussel shells from this area were the result of predation by the crab *Bathogrea thermydron*. Because the effectiveness of this predator declines as the mussels approach 2.0 cm in length, this mussel generates its own size refuge from predation by rapidly growing beyond the size preferred by the crab. As with most bivalves, the larger individuals have stronger shells and are more resistant to decapod predation.

Handling time is probably not limiting in the hydrothermal vent crab–mussel system, but in intertidal areas mobile decapod predators probably minimize handling time rather than maximize food energy (Hughes and Seed 1981; Boulding 1984; Cresswell and McClay 1990). It is not uncommon for bivalve beds, mussels, and oysters, to form in the intertidal zone where predators have limited foraging time and therefore become ineffective at reducing prey population size (Paine 1976). In contrast, Robles (1987) found that the major predator of mussels in the intertidal zone on the southern California rocky shore was the spiny lobster, *Panulirus interruptus*. This predator prefers mussels and moves rapidly into the upper intertidal zone as the tide rises, where it feeds voraciously and selectively on concentrations of mussels. Not unexpectedly, intertidal mussel beds are not found in this area.

In some areas, starfish are thought to be the most effective predators of bivalve populations. These predators feed both extra- and intra-orally and can attach to several bivalves at once (Paine 1974). Starfish, such as *Astropecten* and *Luidia,* that have short or inflexible arms typically swallow their prey whole, with digestion occurring in the stomach and undigested shells being cast out through the mouth. Starfish with more flexible arms, for example *Asterias*, commonly evert their stomach and gradually engulf the prey. In many cases, these starfish use their arms to hold the starfish next to the bivalve or to increase the gape between the bivalve shells and force their everted stomach through the shell gape into the bivalve, where extraoral digestion begins. There are reports of mussels bound with wire being consumed by *Pisaster ochraceus* in this manner in just a few hours (Feder 1955).

Predator–prey studies of the starfish–bivalve interaction are numerous and extensive because of the ease of observation in the rocky intertidal zone. These studies are part of the foundation of our understanding of the causes of rocky intertidal zonation and provide additional evidence of size and spatial refugia in bivalve populations.

As with other equivalent-sized predators, starfish appear to show size preferences for bivalve prey. For example, *Asterias vulgaris* appears to favor small mussels, while *Pisaster ochraceus* feeding on *Mytilus californianus* may take all sizes but favor smaller sizes (Paine 1976). As noted for other predators, predator preference for smaller size classes of bivalves leads to a size refuge for large bivalves and is favored by high bivalve growth rates (Paine 1974, 1976).

Spatial refugia are commonly found in rocky intertidal zone bivalve populations experiencing starfish predation. In most of these cases, the starfish prey heavily on bivalves in the lower portions of the intertidal zone and down into the subtidal zone. The middle portions of the intertidal zone become a refuge because starfish predation is less effective there. This reduced predation effectiveness is thought to be due to the decreased time available for foraging and increased time of exposure to desiccation stress higher in the intertidal zone. Thus, in many temperate rocky intertidal zone habitats, bands of bivalves, mussels and/or oysters, may be found that are limited by physical stress at their upper distribution and predation at their lower limit in the intertidal zone (Menge 1992).

In some subtidal environments, starfish may also control the occurrence of bivalves. Himmelman and Dutil (1991) found that in the Gulf of St. Lawrence the lower limit of dense subtidal beds of *Mytilus edulis* was controlled by the starfishes *Asterias vulgaris* and *Leptasterias polaris*. At greater depths, *Leptasterias* appears to use its tube feet to dig out other species of large infaunal bivalves.

The most mobile and ubiquitous predators in the marine environment are the fishes. However, few studies have focused on the predation of bivalves by fishes. Of the cartilaginous fishes, only the rays are reported to prey on bivalves. Along the Atlantic coast of North America, the cow-nosed ray, *Rhinoptera bonasus,* is reported to feed not only on clams and other bivalves in soft sediment, but also on adult oysters (Smith and Merriner 1978).

Although much more numerous than cartilaginous fishes, little is known about bivalve predation by boney fishes. From the mid-Atlantic to Gulf coasts of North America, the black drum, *Pogonias cromis*, is known to feed on postsettlement oysters (Galtsoff 1964). Schools of this fish invade oyster reefs, where they use their powerful pharyngeal teeth to crush shells and devour the bivalve. In the rocky intertidal habitats of New England, the cunner, *Tautogolabrus adspersus*, is a major predator on *Mytilus* (Olla et al. 1975). The importance of this labrid fish in structuring intertidal communities has been touted to be at least as great as that of the snail *Thais lapillus* (Edwards et al. 1982). Likewise, two labrids, the rock wrasse (*Halichoeres semisinctus*) and the sheephead (*Semicossyphus pulcher*), are thought to be significant consumers of rocky intertidal mussels on Southern California shores (Robles 1987).

Fish are a major cause of a sublethal form of predation known as siphon nipping. For a variety of clams, portions of the exposed siphon are regularly eaten by fishes and occasionally by birds. These losses may significantly reduce clam growth (Coen and Heck 1991). Thus, sublethal predation of siphon tissue can be a major food source for fishes (Trevallion et al. 1971; DeVlas 1979, 1985; Peterson and Quammen 1982). For example, in Rhoad River estuary of Chesapeake Bay, the sciaenid fish, *Leiostomous xanthurus*, and the sole, *Trinectes maculatus*, nipped large numbers of *Macoma balthica* siphons as a major source of food (Haines et al. 1990).

Coastal waters are important feeding areas for many species of birds. Although some birds confine their bivalve predation to rocky intertidal areas, i.e., turnstones and sandpipers, others focus on soft bottoms, i.e., ducks, and still others are found in most shallow water and intertidal habitats, i.e., waders and gulls (Marsh 1986; Meire 1993). Birds have essentially three methods of feeding on bivalves: (1) they may swallow the shell whole, (2) they may remove the flesh from the shell, or (3) they may rob the prey from another bird (Meire 1993). If the shell is swallowed whole, it is crushed in the bird's gizzard and the shell fragments are regurgitated as pellets or defecated. When the flesh is removed, as is the case with oystercatchers, the bill is stabbed between the gaping valves or the shell is hammered until it breaks. Once the bill breaches the shell, the adductor muscle is cut and the valves are pried apart by opening the bill inside the shell. Birds with chisel-shaped bills are mainly stabbers and those with blunt bills are hammerers. Some birds, e.g., gulls and crows, may use another unique method to remove the flesh. The prey is taken from the intertidal habitat and after the bird reaches a suitable altitude the bivalve is repeatedly dropped onto a hard surface until the shell breaks (Meire 1993).

Meire (1993) has synthesized the role of birds in bivalve predation on shallow and intertidal soft sediments. He identifies a number of bivalve characteristics that should enhance survival in these habitats: (1) a high growth rate allows the bivalve to grow into a size refuge; (2) for infauna bivalves, increasing burial depth reduces the probability of being preyed upon and, thus, provides a spatial refuge; and (3) a high reproductive rate assures that the bivalve population will recruit more individuals than will be preyed upon by birds.

In rocky intertidal habitats in Oregon, Marsh (1986) showed that surfbirds (*Aphriza virgata*), gulls (*Larus glauescens*), and black oystercatchers (*Haematopus bachmani*) acting in concert decreased the density of mussels. The intensity of this predation was a function of mussel density and substrate relief, where mussels on vertical walls were unavailable to the birds. Marsh (1986) concluded that a multiple predator assemblage can be effective because its members are active in different microhabitats or prey on different size mussels.

Birds can interact with other predators to structure the populations of intertidal bivalves. Sanchez-Salazar et al. (1987) and Griffiths (1990) found that the shore crab, *Carcinus maenas*, and oystercatchers, *Haematopus ostralegus*, could control the size and distribution of cockles, *Cerastoderma edule*, in the intertidal zone. Cockles settled at relatively low densities throughout the intertidal zone, with those at higher elevations growing more slowly. Although space was not limiting, the shore crabs fed voraciously on small cockles low on the shore and oystercatchers ate large cockles high on the shore. The resulting distribution of cockles was spat and a few large cockles near the low tide mark and small bivalves higher in the intertidal zone (Figures 25 and 26).

In shallow water areas, flocks of birds can decimate natural and cultivated bivalve populations. In Olympia Bay of Puget Sound, Washington, several species of duck utilize oysters as a major food and, as a result, severely damaged the oyster industry (Galtsoff 1964). In Europe and the British Isles, immense flocks of eider ducks (*Somateria*) and oystercatchers (*Haematopus*) can produce heavy bivalve mortalities in wave-protected environments. In Holland, Craeymeersch et al. (1986) reports that available mussel production appears to be the limiting factor on over-wintering oystercatchers, and Meire and Ervynck (1986) have estimated that these birds may consume almost half of the annual mussel production.

Marine mammals are primarily observed at the sea surface, but their feeding activities normally occur underwater beyond the view of observers (Oliver et al. 1983). Of the marine mammals, only a few, sea otters, bearded seals, and walrus, prey on bivalve communities and have been studied in any detail.

With the decline in the fur trade, sea otter populations have gradually been reestablished both artificially and naturally in the cold-temperate nearshore waters of the North Pacific Ocean from Hokkaido to the California peninsula (Van Blaricom 1988). With this re-establishment, the role of sea otters in benthic predation and ecosystem dynamics has come under close scrutiny.

Sea otters forage on a large variety of the benthos of rocky and soft-sediment systems. Prey are dislodged from rocks or dug from the sediments by the forepaws

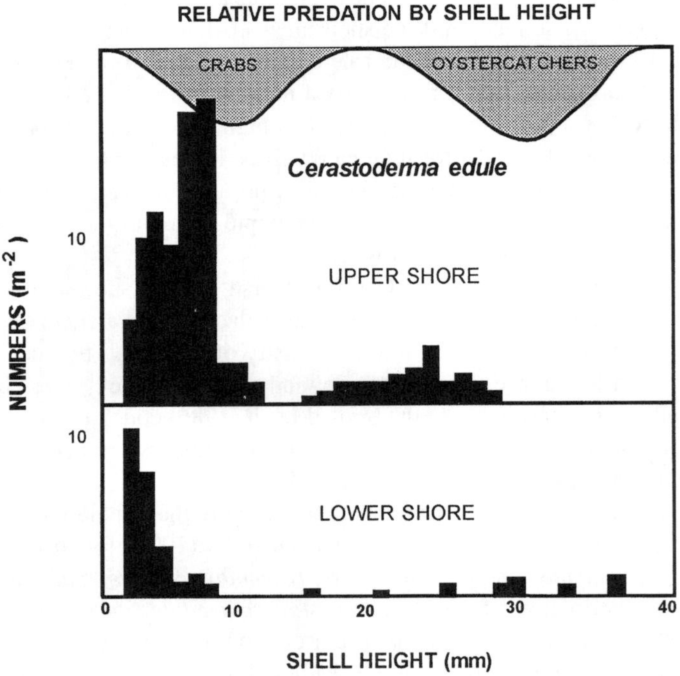

FIGURE 25. The spatial variation in the predation of cockles by crabs and oystercatchers (Adapted from Griffiths, C.L. 1990.)

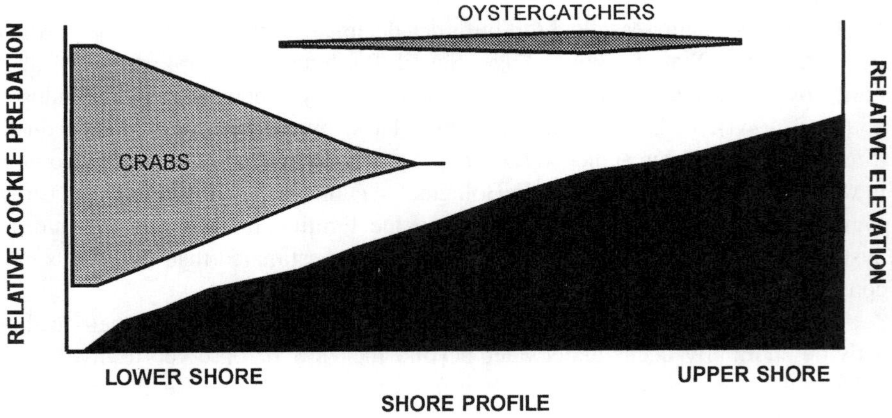

FIGURE 26. The relative predation intensity of crabs and oystercatchers on cockles across and intertidal gradient (Adapted from Griffiths, C.L. 1990.)

and carried to the surface where they are eaten. The shells of bivalves are typically cracked with rocks or other hard objects and then consumed (Riedman and Estes 1988). Their diet is quite broad and depends on the abundance of prey.

On rocky shores, sea otters have been reported to prey on both subtidal and intertidal mussel populations, occasionally creating gaps in the mussel-dominated zone. Because of the impact of their predation on rocky intertidal communities, sea otters have been called "keystone" predators. Although sea otters are not size selective in the mussels taken from the rocks, they are selective in the mussels they consume. Those not consumed are cast off into deep water where mortality is thought to be total. Mussels clearly do not attain a refuge in size from sea otter predation (Van Blaricom 1988).

In soft bottom areas, sea otters feed on three distinct groups, epifauna, shallow-burrowing infauna, and deep-burrowing infauna. Of the 30 species of prey recorded in these habitats, most were clams (Kvitek and Oliver 1988; Kvitek et al. 1988). Only the deep-burrowing clams seem to have spatial refuge from the sea otters, since the otters must excavate large amounts of sediment at depth to capture these prey (Kvitek et al. 1988). Because rocky habitats are like small islands in a sea of sediment, sea otters may depend on the prey in soft-bottom habitats as much or more than those on the rocky shore (Kvitek and Oliver 1988).

The largest predator of bivalves is the walrus, *Odobenus rosmarus*. These large consumers are common and conspicuous components of the marine mammal fauna inhabiting the Arctic and subarctic waters of the Northern Hemisphere. Walruses are the most abundant bottom-feeding marine mammal and are particularly abundant in the Bering and Chukchi Seas, where well over 200,000 feed upon the bivalve communities found on the Beringian platform (Oliver et al. 1983). Walruses excavate bivalves from soft sediments by sucking and expelling water and by movements of the snout and vibrissae (the tusks are not used in feeding). This feeding process produces pits and furrows in the soft sediment that can be analyzed by scuba divers (Oliver et al. 1985). Once the bivalves are out of the sediments and before consumption, the soft parts are removed by suction from between the shells, and the unbroken shells are discarded to the sea floor (Oliver et al. 1983). In different parts of their range, walruses compete with bearded seals and sea otters for bivalves, but little is known of these relationships (Oliver et al. 1983; Oliver et al. 1985). Because of their abundance and large size, walruses disrupt large areas of the soft sediment bivalve community. Most of the bivalves consumed have a generation time of about 3 years. Therefore, it has been hypothesized that the walrus population may decline or crash because the prey population cannot regenerate fast enough to sustain the walruses. Since these large bivalve predators are seldom hunted by humans, time should answer this classical predator–prey cycle question (Oliver et al. 1985).

Many other predators have been reported for bivalve populations. Sea turtles and octopi are predators of *Tridacnid* clams in the South Pacific (Copland and Lucus 1988). Diamondback terrapins (Hurd et al. 1979), and sea urchins (Seed 1992) are thought to prey on *Mytilus* in different environments. Without a doubt, there are many, some probably significant bivalve predators that we have missed in this review. Clearly bivalves are a significant food resource to a great variety of marine organisms.

COMPETITION-INDUCED MORTALITY

As discussed earlier, postsettlement bivalves can and do remove the planktonic larvae of their own and other species from the water column. While this form of adult feeding and larval mortality appears incidental, it is also self-serving to the adult, because it removes a potential competitor for food, space, and other resources. More directly, along the west coast of North America, *Mytilus californianus* competes with and dominates *Mytilus edulis,* setting the lower limits of *M. edulis* in exposed rocky intertidal habitats (Suchanek 1978, 1986). Harger (1972) has proposed that *M. californianus* crushes *M. edulis*, but Suchanek (1978) has argued to the contrary and there is no consensus on this idea. In soft sediments, competitive exclusion in space is rare, although competition for food probably exists (Peterson 1991).

AGGREGATED DISTRIBUTIONS

Many bivalves, e.g., oysters, mussels, clams, cockles, etc., form dense assemblages or populations, and these are often called beds or, in the most extreme case, reefs. A reef is defined as an elevated ridge or block of material that may be derived from the skeletal material of organisms, geological substrates, or man-made structures (Hughes 1991). These structures influence local topographic complexity on a large scale, and in so doing influence hydrographic conditions and provide numerous microhabitats for a variety of organisms. Thus, reefs are associated with an increase in biodiversity or species richness when compared to the adjacent environment. While both mussels and oysters form extensive layers of shell on the bottom of estuaries, these layers tend to become unstable upon reaching a critical thickness and collapse as a result of currents and waves (Seed 1969). In the southeastern U.S., the shells of *Crassostrea virginica* can accumulate and form extensive reefs that range from 10 to 1000s of meters in length (Stenzel 1971; Dame 1976).

In New England, the mussel, *Geukensia demissa*, forms dense bands on the edge of *Spartina alterniflora* salt marsh in the mid-intertidal zone. Bertness and Grosholz (1985) found that the effects of conspecifics on the success of individual mussels included reduced growth rates and reduced fecundity. Smaller mussels were affected by crowding more than larger mussels. In contrast, the mortality agents of sediment suffocation, winter ice dislodgment, and predation were all reduced by group living. These investigators suggest that both on hard substrates and soft sediments aggregations of mussels directly affect the most critical variables limiting their populations, i.e., space and sediment stability.

The reduction in population density by competitively induced losses of individuals within a cohort is known as self-thinning or density-dependent disappearance (Hughes and Griffiths 1988). Accordingly, when space is used fully and surviving individuals are still able to grow, the area occupied per individual must be inversely proportional to population density. As weight is proportional to the cube of length, and area occupied by an individual is proportional to the square of its length, this accounts for the exponent, 3/2. These authors were able to show that for the mussel *Choromytilus meridionalis* deviations from this exponent could be accounted for by allometric growth and multilayered packing. Allometric growth is a factor because mussels change their shape as they grow and mature, and layering causes deviations because the mussels in each layer inhibit each other's growth. Thus, self-thinning appears to be a fundamental property of dense bivalve populations.

At a somewhat different spatial scale, the distribution of clumps of large and small oysters on oyster reefs in the Copano Bay-Aransas Bay area along the Texas coast were investigated by Powell et al. (1987). They found that small-scale spatial variation (about 40 cm) in small oysters (<2 cm) was patchily distributed, while large oysters (>5 cm) were less contagiously distributed. Negative spatial autocorrelation was restricted to adjacent clumps <12 cm apart. Thus as the oyster populations aged, their association changed from gregarious (positive) to negative. These investigators speculate that mortality produced by predation and disease, modulated by competition for food, allowed large oysters to affect the survivorship of oysters in adjacent clumps. In these small-scale changes in spatial distribution, beds of bivalves have many of the ecological advantages of clonal style living, good spatial competition with other species, and reduced mortality from both physical and biological agents. The benefits of this life style are greater than the costs of solitary existence.

The dominant bivalves in reefs and beds provide a vertical aspect to the three-dimensional nature of these communities. This structure provides numerous microhabitats for both motile species and other sessile forms (Dean 1981). On intertidal oyster reefs in South Carolina, Dame (1979) found 37 species of macroinvertebrates (Table 10), and this value compares favorably with 42 species found by Bahr and Lanier (1981) on oyster rocks in Georgia. Both systems had Shannon diversity indices of 1.5 to 2.2. Most of the species in these systems are heterotrophic and are dominated functionally by filter feeders, deposit feeders, and predators. In nearby North Carolina, Wells (1961) found 303 species on both intertidal and subtidal oyster reefs along a salinity gradient in the Newport River estuary. Similar efforts on mussel beds (Lintas and Seed 1994; Seed 1996) have shown that like oyster reefs, mussel communities on rocky shores are typically dominated by a few very abundant macroinvertebrate species. These investigators found that *Mytilus edulis* beds on rocky wave-exposed shores in North Wales could be stratified into three layers, each dominated by a single species. A foraminiferan dominated the sediments of the bottom layer, a barnacle was only found on the outer surface of the mussel shells, and the brooding bivalve, *Lasaea rubra*, was only found in the matrix of shell and byssal threads in the middle layer. Shannon diversity varied from 1.7 to 2.8 with highest values low in the intertidal zone. Thus, these systems are clearly diverse in macro-invertebrates and not simple monospecific stands of bivalves.

Following the ideas of island biogeography (MacArthur and Wilson 1967), the number of species in a community increases with the size of the island, because environmental heterogeneity increases with island size. Recent studies by Tsuchiya and Nishihira (1985, 1986) investigated these species–area relationships for patches of *M. edulis* ranging in size from 0.4 to 521.3 cm^2. They found that with increasing patch size, species richness and the number of individuals of associated animals increased, but decreased per unit patch size. Large individuals of some species were only found on the larger patches. In the larger patches, the edges of the patch supported a different environment from the central part. The amounts of sediments,

TABLE 10
Species Composition of Intertidal Oyster Reefs in North Inlet, SC

Species	Ave. No. m^{-2}
Polychaeta	
Amphitrite ornata	15
Cirratulus grandis	10
Glycera americana	4
Heteromastus filiformis	887
Hydroides dianthus	<1
Laeonereis culveri	<1
Lumbrinereis tenuis	44
Marphysa sanguinea	6
Nereis succinea	135
Paleanotus heteroseta	<1
Panopea bitruncata	<1
Pectineria gouldii	3
Pherusa sp.	<1
Phyllodocea fragilis	10
Polycirrus eximius	<1
Polydora ligni	3
Scoloplos fragilis	9
Streblosoma sp.	<1
Arthropoda	
Anurida maritima	6
Balanus eburneus	36
Eurypanopeus depressus	50
Gammarus palustris	6
Insect pupae	34
Melita nitida	23
Panopeus herbstii	42
Pinnotheres ostreum	<1
Uca pugilator	7
Mollusca	
Brachidontes exustus	704
Chione intapurpurea	<1
Crassostrea virginica	802
Geukensia demissa	7
Ilynassa obsoleta	2
Mercenaria mercenaria	1
Boonia impressa	78
Terebra dislocata	<1
Nematoda	<1
Nermertea	19

After Dame 1979.

shell fragments, and byssal threads were greater in the periphery and central part of adult mussel patches than in young and old mussel patches. Total organic matter increased with the age of the patch. Species richness was higher in older patches, and species diversity was higher in adult mussel patches. These findings were, in general, supported by the studies of Lintas and Seed (1994) and Seed (1996) on mussel beds and patches in North Wales. These authors further speculated that small-scale spatial variations exhibited by the fauna associated with *M. edulis* appear to be related to aerial and tidal exposure, mussel density, the amount of accumulated sediment as well as patch size. Although these investigators focused mainly on the biotic aspects of *Mytilus* islands, they clearly recognize the importance of the interactions between system structure and the physical environment in temporal and spatial development. These observations also show that comparisons of bivalve communities based on samples collected at single points in space and time clearly need to be interpreted with caution (Seed 1996).

ZONATION

The intertidal zone is characterized by vertical gradients in both physical and biotic factors that control populations and communities dominated by bivalves. On rocky coastlines throughout the world, bivalves, particularly mussels in the family Mytilidae, have the ability to dominate portions of the intertidal zone. In addition to being eurytopic, with abilities to withstand extremes of physical factors, i.e., temperature, salinity, desiccation, and oxygen tension, the members of this family develop byssal attachment threads that allow them to exploit hard substrates and dominate rocky habitats (Suchanek 1986). These mussels often form dense bands of animals that occupy intertidal space, particularly in temperate regions (Seed and Suchanek 1992). The majority of Mytilidae can also survive in the subtidal zone.

Numerous studies (see Seed and Suchanek 1992) have shown that the upper limit of mussel bands on rocky intertidal environments is controlled by physical factors, while the lower limit is usually controlled by predation and competition. These general observations support those of Connell (1961, 1972) on intertidal barnacles and the speculation that such zonation control was applicable to other intertidal forms.

Peterson (1991) has argued persuasively that zonation of bivalves and other invertebrates also occurs in intertidal sediments. These environments are typically sand and mud flats that are often quite extensive in shallow coastal systems. As in the rocky intertidal zone, gradients of physical factors lead to gradients of physiological processes. The dominant factors are exposure to air and light, temperature, and wave action. Thus, higher elevations are more rigorous, and only a few species can tolerate these stresses. Predation is frequently evidenced by rapidly moving fishes, crabs, and birds, as well as terrestrial predators. In contrast to the rocky intertidal zone, where competition for space can lead to sharp boundaries between zones, competitive exclusion is rare in these soft sediment systems. The greatly reduced competition for space and the indirect effects of competition for food leads to more gradual zonation dominated by physical effects (Black and Peterson 1988).

Intertidal oyster reefs and mussel beds are usually found in creeks and channels surrounded by soft sediment environments. Mussel and oyster beds are exposed to the same gradients of physical factors as dense populations of bivalves in the rocky intertidal zone and less dense populations in mud and sand flats. As the beds grow vertically, the effects of the physical environment, particularly exposure and temperature, can become acute. These systems appear to suffer predation from both rapidly moving birds, crabs, and fish as well as sluggish forms as gastropods and fungi. There is strong competition for space that is modulated by mussels and oysters growing on top of each other and thus increasing the vertical dimension of their communities. This increase in system structure in the form of more microhabitats leads to a moderation in competitive exclusion and the existence of other bivalves in significant numbers, i.e., *Brachidontes* mussels on oyster reefs (Table 10).

DISEASES AND PARASITES

The healthy state of a bivalve mollusc entails normal structural and functional status, where the animal can successfully accommodate the typical ecological conditions as they influence growth, development, reproduction, survival, and competition (Kinne 1980). Disease is the deviation or reduction (negative deviation) from the healthy state of the living animal. There is a gradient from the healthy to the diseased state.

A disease is demonstrable when the disease-causing agent surpasses certain critical levels. Critical levels can be identified by (1) an increase in the number of a given disease carrying agents per host bivalve; (2) an increase in the agent's virulence or destructive potential; (3) the concomitant activities of several agents of different taxa; (4) a decrease in host resistance; and (5) changes in the quantitative relationship between host and agent (Kinne 1980).

Although diseases can be restricted to an individual bivalve, this book will focus on those that are transmitted from one individual to another, i.e., infectious, communicable, and contagious diseases. Four types of disease can be identified: (1) endemics (enzootics) are diseases that are characteristic of a host population in a certain area; (2) epidemics (epizootics) are temporary mass outbreaks of a communicable disease; (3) pandemics (panzooics) are temporary mass outbreaks of a communicable disease over a wide geographic area; and (4) sporadic (ecdemic) diseases are more or less atypical of the area concerned (Kinne 1980).

Diseases have a number of different causes:

1. Circumstances internal to the individual
2. Nutritional disorders
3. Effects of abiotic physical factors and pollutants
4. Physical injuries
5. Agents or coexisting organisms
6. Combinations of the preceding five causes

This text will focus on the biotic diseases or agents, including viruses, bacteria, fungi, protozoans, mesozoans, and metazoans.

In 1983, Kinne declared that more is known about the diseases of bivalves than of the diseases of all other marine invertebrates combined. Because of the voluminous amount of published literature on marine bivalve diseases, it is impossible to do the topic justice in this short space. However, those interested in detailed accounts are encouraged to see Kinne (1980 and 1983), Sparks (1985), Sinderman (1990), and Perkins (1993) for comprehensive reviews of bivalve diseases.

For the purposes of this text, we will quickly summarize the common disease-causing biotic agents of marine bivalves.

More than 20 viruses have been described for marine bivalves (Sinderman 1990). Few have been adequately characterized and then only with histopathology and ultrastructure morphology. The major problem with viral diseases in shellfish is the demonstration of causal relationships between the virus and the observed pathology. A number of different viral types have been observed in commercially abundant bivalves including herpes virus, reovirus, papovavirus, icosahedral virus, iridovirus, retrovirus, and papillomavirus. Each virus appears to invade a specific tissue with infections found in digestive glands, gonads, gills, palps, and connective tissues. Of these viruses, an epidemic (epizootic) of iridovirus that causes gill disease in *Crassostrea angulata* resulted in the loss of this species to cultivation along the coast of France in the 1960s and 1970s (Sinderman 1990 and Perkins 1993).

Bacterial (Prokaryotes) diseases of bivalves appear to be important in the presettlement stages and are for the most part unimportant in adults. This is probably due to the well-developed defense mechanisms of phagocytosis and encapsulation common in postsettlement bivalves. Successful bivalve bacterial agents appear to be able to avoid or escape these defenses. By far the most important disease-causing bacterium for bivalves is *Vibrio* that can be deadly to developing larvae (Sinderman 1990 and Perkins 1993).

The fungi (Eumycota) are poorly represented in the litany of disease-causing agents of bivalves. The major observed fungal agent is an aquatic phycomycetes that attacks the shell of adult bivalves and is known as "shell disease" or *maladie du pied*. This fungus, *Ostracoblabe implexa*, has been reported to cause mass mortality in both natural and cultivated stocks of European oysters. In this disease, the shell in the area of the muscle attachment is attacked. Eventually the muscle becomes detached and death follows (Sinderman 1990 and Perkins 1993).

The protozoans (Protista) are probably the most common cause of epizootic outbreaks that result in mass mortalities in populations of economically important bivalves. Although protozoan infections have been reported in many bivalves including mussels, clams, and shipworms, it is in the oysters where they have been most extensively studied. The most notable of the many protozoan diseases are dermo, caused by a member of the phylum Apicomplexa, *Perkinsus marinus*, and MSX caused by *Haplosporidium nelsoni* of the phylum Ascetospora (Sparks 1985, Sinderman 1990). Dermo is characterized by emaciation of the digestive gland in oysters. Adult oysters appear to be most susceptible after spawning, and mortality increases with age and size (Lauckner 1983). MSX also attacks and emaciates oysters, and both diseases have limits imposed by salinity and temperature. Both of these protozoans have caused epidemics in natural and cultivated oyster populations along the Gulf and Atlantic coasts of North America, to the extent that oyster harvests in these areas are but a fraction of previous yields. Although protozoan epidemics are reported worldwide, it is thought that these massive population mortalities have always existed. Their spread has been helped by experimental plantings of oysters from other regions. As is the case for most disease, dense populations typical of cultivated plots and growout facilities are most susceptible to disease transmission.

Parasitic worms, trematodes, cestodes, and nematodes can reduce growth and fecundity within bivalves and may in some cases cause death (Lauckner 1983; Sinderman 1990). Polychaete annelid worms of the genus *Polydora* (mud or blister worms) can invade oysters, mussels, and scallops and in cases of severe infection cause death. The larvae of *Polydora* settle on the exterior of bivalves where they proceed to make burrows at the edge of the valves. The worm accumulates mud around itself and the bivalve responds by secreting shell to cover the mud–worm complex. Bivalves heavily infested by *Polydora* are usually of poor condition and weak (Galtsoff 1964; Skeel 1979). Some bivalves may die as a consequence of *Polydora* infestation, but most are weakened both physiologically (condition) and structurally (perforated shells) and thus be more susceptible to other forms of predation. In a study of *Polydora ciliata* infestations of *Mytilus edulis*, Ambariyanto and Seed (1991) found that there was a higher incidence of these parasites in larger mussels and that heavily infested mussels were more vulnerable to predation by shore crabs. These investigators also found that *P. ciliata* infestations were highest at the mouth of the estuary and declined along a gradient up the estuary. Unlike other parasites of bivalves, these shell-borers showed no systematic variation with tidal elevation.

Using Taylor's (1984) general definition of predation, the small pyramidellid gastropods can be considered predators or parasites. Such a gastropod, *Boonea impressa*, is a common member of *Crassostrea virginica* dominated oyster reefs in the southern U.S. This gastropod feeds, as do other pyramidellids, by attaching to the soft tissues of the bivalve with a sucker, then piercing the tissues with a buccal stylet and sucking the body fluids of the prey (Fretter and Graham 1949; Allen 1958). Small *Boonea* may be a direct cause of oyster spat mortality (White et al. 1988), while the valve movements, filtration rate, and growth rate of larger bivalves are affected by the parasite (Ward and Langdon 1986). Using an energy flow model, White et al. (1988) determined that these snails appear to have less impact on populations of large oysters, and because of *Boonea's* patchy distribution, its impact could not be uniform. Further work by this group showed that *Boonea* can be a vector for the transmission of dermo disease, since individuals migrate between oysters.

Parasitic crustaceans that are common in bivalves often destroy the gonads, influence nutrient adsorption and modify or inhibit gill function (Sinderman 1990). A number of species of the parasitic copepod *Mytilicolas* infect the digestive track of different bivalves including mussels, oysters, and clams. The well-known pea crab, *Pinnotheres ostreum*, inhabits the mantle cavity of *Crassostrea virginica,* and other pinnotherids parasitize scallops and mussels. Developing pea crabs invade the bivalve, where the crabs mature to sexually dimorphic forms of relatively large females and tiny males. Originally, these crustaceans were thought not to harm the bivalve (Galtsoff 1964), but the current belief is that pea crabs interfere with growth by eroding the gills (Kruczynski 1972; Bierbaum and Ferson 1986; Bierbaum and Shumway

1988). In oysters, it is thought that pea crabs may influence sex ratio in favor of male bivalves (Awati and Rai 1931; Christensen and McDermott 1958), and this ratio change could have population effects. Although pea crabs are usually categorized as parasites or pests, using the more general definition of predation they can and do reduce the productivity of bivalves at both the individual and population levels.

Disease is an important ecological factor. At the individual level, disease can affect the rates and efficiencies of metabolic processes including, growth, filtration, respiration, and reproduction. It may change the timing of the life cycle and the capacities for regulation and adaptation of the host bivalve to stress. Disease may also change the structural morphology of not only the soft body parts, but also the shell. Ultimately disease reduces the life span of the organism and its success in its environment.

At the population level, epidemics can change population structure and reduce population size. Population distribution and abundance can be influenced both spatially and temporally and population reproduction is often reduced. If critical levels are reached, entire populations may go extinct.

From an ecosystem perspective, epidemic diseases will directly influence the functional status of the host bivalve and indirectly radiate to the rest of the system through the food web. The loss of an entire functional component to disease, i.e., filter feeders, can change the entire ecosystem.

POPULATION ENERGY BUDGETS

As is presented in Chapter 3, Organismic Level Processes, the components of an organism's energy budget can be determined by examining the feeding, respiration, excretion, fecal production, growth, and reproduction of bivalves of different sizes. These values can be scaled-up to population patches or entire populations by taking into account natality and mortality patterns and integrating estimates of population density, usually in the form of the age or size histograms described earlier. Basically, this scaling-up is an integration of laboratory measurements of organismic processes with field measurements of actual populations.

In scaling up to the population level, care should be taken to estimate or examine two components of the budget (Griffiths and Griffiths 1987). First, respiration (R) as measured in the laboratory is often underestimated, due to the constraints of the respirometer on the individual bivalve. Although it is not normally a problem with bivalves, Trevallion (1971) advocates multiplying laboratory measurements by a factor of 2 to better approximate actual field respiration. Second, and of more concern, laboratory estimates of consumption (C) are usually made with monospecific cultures of phytoplankton. In nature, bivalves feed on water that contains not only many species of algae, but also suspended organic and inorganic particles. Thus the quality of the natural food may be very different from the laboratory ration (Stuart 1982). To circumvent these problems, Crosby (personal communication) has developed a method to measure feeding, excretion, and respiration in the field in flow-through chambers, but a rigorous comparison of this technique with the integrated field and laboratory approach has yet to be done.

Population energy budgets are usually used to compare different populations or the same population in different years. These budgets are acutely sensitive to changes in size and age composition as well as changing environmental conditions (Griffiths and Griffiths 1987). In spite of the difficulties raised earlier, it is still worthwhile to compare the energy budgets of different bivalve populations (Table 11).

From Table 11, it is obvious that denser populations consume more energy than less dense populations. Most bivalve populations put less than 10% of their intake into growth and slightly more than 10% of their intake into growth and reproduction combined. Most consumption goes into respiration and waste. The production to biomass ratio (P/B) is an estimate of the turnover of energy in the population. The P/B ratio averages to 1.2, but this ratio is very sensitive to size distribution, with small fast-growing size classes having higher P/B ratios and older size classes having smaller ratios (Griffiths and King 1979).

Clearly, there are few generalities regarding bivalve population energy budgets. Although they give a much more detailed view of the functional aspects of a given bivalve population, each population should be considered a unique case. When the informational need is apparent and the resources are available, then population specific energy budgets should be constructed.

TABLE 11
Population Energy Budgets for Various Bivalve Species, Expressed as a Percentage of Consumption (C)

Species (%)	C (kJ m^{-2} year^{-1})	P_g (%)	P_r	R (%)	F + U (%)	P/B ratio	Ref.
Aulacomya ater	27,499	3.4	5.8	49.3	41.5	1.3	Griffiths and King (1979)
Cardium edule	3,622	2.8	—	57.2	40.0	—	Loo and Rosenberg (1989)
Chlamys islandica	55,711	1.0	0.4	3.6	94.2	0.76	Vahl (1981)
Choromytilus meridionalis	831,890	1.6	4.8	10.4	83.2	2.91	Griffiths (1981)
Crassostrea gigas	10,140	0.4	20.7	20.7	63.7	—	Bernard (1974)
Crassostrea virginica	8,811	17.7	3.4	28,9	50.0	2.0	Dame (1976)
Geukensia demissa	468	12.4	2.5	34.8	50.0	0.28	Kuenzler (1961)
Lyrodus pedicellatus	—	9.0	4.3	19.8	66.7	—	Gallager et al. (1981)
Mercenaria mercenaria	5,400	5.6	4.7	18.7	71.1	0.39	Hibbert (1977)
Mya arenaria	545	6.8	—	53.2	40.2	—	Loo and Rosenberg (1989)
Mytilus edulis	—	8.9	4.8	25.8	60.4	—	Bayne and Newell (1983)
Ostrea edulis	468	6.4	5.5	29.0	59.1	—	Rodhouse (1979)
Patinopecten yessoensis	—	22.7	4.3	43.6	29.4	—	Fuji and Hashizume (1974)
Scrobicularia plana	2,315	6.6	6.2	48.0	39.3	—	Hughes (1970)
Tellina tenuis	703	1.2	0.9	7.9	90	—	Trevallion (1971)

Note: C = Consumption; P_g = production as growth; P_r = production as reproduction; and F + U = feces and urine (excreted waste).

After Griffiths and Griffiths (1987).

REFERENCES

Allen, J. 1958. Feeding habits of two species of *Odostomia*, *Nautilus*, 72, 11–15.

Ambariyanto and Seed, R. 1991. The infestation of *Mytilus edulis* Linnaeus by *Polydora ciliata* (Johnston) in the Conwy estuary, North Wales, *J. Moll. Stud.*, 57, 413–424.

Andre, C. and Rosenberg, R. 1991. Adult–larval interactions in the suspension feeding bivalves *Cerastoderma edule* and *Mya arenaria*, *Mar. Ecol. Prog. Ser.*, 71, 227–234.

Andre, C., Jonsson, P.R., and Lindegarth, M. 1993. Predation on settling bivalve larvae by benthic suspension in feeders: the role of hydrodynamics and larval behaviour, *Mar. Ecol. Prog. Ser.*, 97, 183–192.

Awati, P.R. and Rai, H.S. 1931. *Ostrea cucullata* (the Bombay oyster), *Indian Zool. Mem.*, 3, 1–107.

Bahr, L.M. and Lanier, W.P. 1981. The ecology of intertidal oyster reefs of the South Atlantic coast: a community profile, FWS/OBS-81/15, U.S. Fish and Wildlife Service, 105 pp.

Barnes, R.D. 1974. *Invertebrate Zoology*, 3rd ed., Saunders, Philadelphia, 870 pp.

Basson, M., Rosenberg, A.A., and Beddington, J.R. 1988. The accuracy and reliability of two new methods for estimating growth parameters from length-frequency data, *J. Cons. Int. Explor. Mer*, 44, 277–285.

Bayne, B.L. 1976. Aspects of reproduction in bivalve molluscs, in Wiley, M., Ed., *Estuarine Processes*, Vol. 1, *Uses, Stresses and Adaptation to the Estuary*, Academic Press, New York, 432–448.

Bayne, B.L. and Newell, R.C. 1983. Physiological energetics of marine molluscs, in Wilbur, K.M. and Saleuddin, A.S.M., Eds., *The Biology of Molluscs: Physiology*, Academic Press, New York, 407–515.

Bernard, F.R. 1974. Annual biodeposition and gross energy budget of mature Pacific oysters *Crassostrea gigas*, *J. Fish. Res. Board Can.*, 31, 185–190.

Berry, P.F. 1978. Reproduction, growth and production in the mussel, *Perna perna* (Linnaeus), on the east coast of South Africa, *Invest. Rep. Oceanogr. Res. Inst. Durban*, 48, 1–28.

Bertness, M.D. and Grosholz, E. 1985. Population dynamics of the ribbed mussel, *Geukensia demissa*: the costs and benefits of an aggregated distribution, *Oecologia*, 67, 192–204.

Bierbaum, R.M. and Ferson, S. 1986. Do symbiotic pea crabs decrease growth rate in mussels?, *Biol. Bull.*, 170, 51–61.

Bierbaum, R.M. and Shumway, S. 1988. Filtration and oxygen consumption in mussels, *Mytilus edulis*, with and without pea crabs, *Pinnotheres maculatus*, *Estuaries*, 11, 264–271.

Black, R. and Peterson, C.H. 1988. Absence of preemption and interference competition for space between large suspension-feeding bivalves and smaller infaunal macroinvertebrates, *J. Exp. Mar. Biol. Ecol.*, 120, 183–198.

Boulding, E.G. 1984. Crab-resistant features of shells of burrowing bivalves: decreasing vulnerability by increasing handling time, *J. Exp. Mar. Biol. Ecol.*, 76, 201–223.

Brousseau, D.J. 1978. Population dynamics of the soft-shell clam *Mya arenaria*, *Mar. Biol.*, 50, 63–71.

Carpenter, S.R. and Kitchell, J.F. 1993. *The Trophic Cascade in Lakes*, Cambridge University Press, Cambridge, 385 pp.

Cargo, D.G. and Shultz, L.P. 1967. Further observations on the biology of the sea nettle and jellyfishes in Chesapeake Bay, *Chesapeake Sci.*, 8, 209–220.

Cassie, R.M. 1954. Some uses of probability paper in the analysis of size frequency distributions, *Aust. J. Mar. Freshw. Res.*, 5, 513–522.

Cerrato, R.M. 1980. Demographic analysis of bivalve populations, in Rhoads, D.C. and Lutz, R.A., Eds., *Skeletal Growth of Aquatic Organisms*, Plenum Press, New York, 417–468.

Christensen, A.M. and McDermott, J.J. 1958. Life-history and biology of the oyster crab, *Pinnotheres ostreum* (Say), *Biol. Bull.*, 114, 146–179.

Coen, L.D. and Heck, K.L. 1991. The interacting effects of siphon nipping and habitat on bivalve (*Mercenaria mercenaria* L.) growth in a subtropical seagrass (*Halodule wrightii* Aschers) meadow, *J. Exp. Mar. Biol. Ecol.*, 145, 1–13.

Connell, J.H. 1961. Effects of competition, predation by *Thais lapillus*, and other factors on natural populations of the barnacle *Balanus balanoides*, *Ecol. Monogr.*, 31, 61–104.

Connell, J.H. 1972. Community interactions on marine rocky intertidal shores, *Annu. Rev. Ecol. Syst.*, 3, 169–192.

Copland, J.W. and Lucas, J.S. 1988. *Giant Clams in Asia and the Pacific*, Australian Center for International Agricultural Research, Canberra, 274 pp.

Craeymeersch, J.A., Herman, P.M.J. and Meire, P.M. 1986. Secondary production of an intertidal mussel (*Mytilus edulis* L.) population in the Eastern Scheldt, *Hydrobiologia*, 133, 107–115.

Cresswell, P.D. and McClay, C.L. 1990. Handling times, prey size and species selection by *Cancer novaezelandiae* feeding on molluscan prey, *J. Exp. Mar. Biol. Ecol.*, 140, 13–29.

Dame, R.F. 1976. Energy flow in an intertidal oyster population, *Estuarine Coastal Mar. Sci.*, 4, 243–253.

Dame, R.F. 1979. The abundance, diversity and biomass of macrobenthos on North Inlet, South Carolina, intertidal oyster reefs, *Proc. Natl. Shellfish. Assoc.*, 69, 6–10.

Dayton, P.K. 1971. Competition, disturbance, and community organization: the provision and subsequent utilization of space in a rocky intertidal community, *Ecol. Monogr.*, 41, 351–389.

Dean, T.A. 1981. Structural aspects of sessile invertebrates as organizing forces in an estuarine fouling community, *J. Exp. Mar. Biol. Ecol.*, 53, 163–180.

DeVlas, J. 1979. Annual food intake by plaice and founder in a tidal flat area in the Dutch Wadden Sea, with special reference to consumption of regenerating parts of macrobenthic prey, *Neth. J. Sea Res.*, 13, 117–153.

DeVlas, J. 1985. Secondary production by siphon regeneration in a tidal flat population of *Macoma balthica*, *Neth. J. Sea Res.*, 19, 147–164.

Edwards, D.C., Conover, D.O., and Sutter, F. 1982. Mobile predators and the structure of marine intertidal communities, *Ecology*, 63, 1175–1180.

Feder, H.M. 1955. On the methods used by the starfish *Pisaster ochraceus* in opening three types of bivalve molluscs, *Ecology*, 36, 764–767.

Fisher, R.A. 1958. *The Genetical Theory of Natural Selection*, Dover, New York, 291 pp.

Fretter, V. and Graham, A. 1949. The structure and mode of life of the Pyramidellidae, parasitic opisthobranch, *J. Mar. Biol. Assoc. U.K.*, 28, 493–532.

Fuji, A. and Hashizume, M. 1974. Energy budget for a Japanese common scallop, *Patinopecten yessoensis* (Jay), in Mutsu Bay, *Bull. Fac. Fish. Hokkaido Univ.*, 25, 7–19.

Gallager, S.M., Turner, R.D., and Berg, C.J. 1987. Physiological aspects of wood consumption, growth, and reproduction in the shipworm *Lyrodus pedicellatus* Quatrefages (Bivalvia: Teredinidae), *J. Exp. Mar. Biol. Ecol.*, 52, 63–77.

Galleni, L., Ferro, E., Salghetti, V., Tongiorgi, P. and Salvadego, P. 1977. Ulteriori osservazioni sulla predazione di *Stylochus mediterraneus* sui Mytili e suo orientamento chemiotattico, *Atti IX Congr. Soc. Ital. Biol. Mar. Lacco Ameno d'Ischia*, pp. 259–261.

Galtsoff, P.S. 1964. The American oyster *Crassostrea virginica* Gmelin, *Fish. Bull.*, 64, 1–480.

Griffiths, C.L. 1990. Spatial gradients in predation pressure and their influence on the dynamics of two littoral bivalve populations, in Morton, B., Ed., *The Bivalvia — Proc. Mem. Symp. in Honour of Sir Charles Maurice Yonge, Edinburgh, 1986*, Hong Kong University Press, Hong Kong, 321–332.

Griffiths, C.L. and Griffiths, R.J. 1987. Bivalvia, in Pandian, T.J. and Vernberg, F.J., Eds., *Animal Energetics*, Vol. 2, *Bivalvia through Reptilia*, Academic Press, New York, 1–88.
Griffiths, C.L. and King, J.A. 1979. Energy expended on growth and gonad output in the ribbed mussel *Aulacomya ater*, *Mar. Biol.*, 53, 217–222.
Griffiths, R.J. 1981. Production and energy flow in relation to age and shore level in the bivalve *Choromytilus meridionalis*, *Estuarine Coastal Mar. Sci.*, 12, 101–118.
Haines, A.H., Haddon, A.M. and Wiechert, L.A. 1990. Guild structure and foraging impact of blue crabs and epibenthic fish in a sub-estuary of Chesapeake Bay, *Mar. Ecol. Prog. Ser.*, 67, 105–126.
Harger, J.R.E. 1972. Competitive co-existence: maintenance of interacting associations of the sea mussels *Mytilus edulis* and *Mytilus californianus*, *Veliger*, 14, 387–410.
Hibbert, C.J. 1977. Energy relations of the bivalve *Mercenaria mercenaria* on an intertidal mudflat, *Mar. Biol.*, 44, 77–84.
Himmelman, J.H. and Dutil, C. 1991. Distribution, population structure and feeding of subtidal seastars in the northern Gulf of St. Lawrence, *Mar. Ecol. Prog. Ser.*, 76, 61–72.
Hughes, R.N. 1970. An energy budget for a tidal-flat population of the bivalve *Scrobicularia plana* (Da Costa), *J. Anim. Ecol.*, 39, 357–381.
Hughes, R.N. 1991. Reefs, in Barnes, R.S.K. and Mann, K.H., Eds., *Fundamentals of Aquatic Ecology*, 2nd ed., Blackwell Scientific, Oxford, 213–229.
Hughes, R.N. and Griffiths, C.L. 1988. Self-thinning and mussels: the geometry of packing, *Am. Nat.*, 132, 484–491.
Hughes, R.N. and Seed, R. 1981. Size selection of mussels by the blue crab *Callinectes sapidus*: energy maximizer or time minimizer?, *Mar. Ecol. Prog. Ser.*, 6, 83–89.
Hurd, L.E., Smedes, G.W., and Dean, T.A. 1979. An ecological study of a natural population of diamondback terrapins (*Malaclemys t. terrapin*) in a Delaware salt marsh, *Estuaries*, 2, 28–33.
Kinne, O. 1980. Diseases of marine animals: general aspects, in Kinne, O., Ed., *Diseases of Marine Animals*, Vol. 1, John Wiley, New York, 13–74.
Kinne, O., Ed., 1983. *Diseases of Marine Animals*, Vol. 2, Biologische Anstalt. Helgoland, 961 pp.
Korringa, P. 1949. *Crepidula fornicata* as an oyster-pest, *J. Cons. Perm. Int. Explor. Mer., Rapp. P.-V.Reun.*, 128, 55–59.
Krebs, C.H. 1978. Ecology: *The Experimental Analysis of Distribution and Abundance*, 2nd ed., Harper and Row, New York, 678 pp.
Kremmer, P. 1979. Predation by the ctenophore *Mnemiopsis leidyi* in Narragansett Bay, Rhode Island, *Estuaries*, 2, 97–105.
Kruczynski, W.L. 1972. Distribution and abundance of *Pinnotheres maculatus* Say in Bogue Sound, North Carolina, *Biol. Bull.*, 145, 482–491.
Kuenzler, E.J. 1961. Structure and energy flow of a mussel population in a Georgia salt marsh, *Limnol. Oceanogr.*, 6, 191–204.
Kurten, B. 1964. Population structure in paleoecology, in Imbrie, J. and Newell, N., Eds., *Approaches to Paleoecology*, John Wiley, New York, 91–106.
Kvitek, R.G., Fukayama, A.K., Anderson, B.S., and Grimm, B.K. 1988. Sea otter foraging on deep-burrowing bivalves in a California coastal lagoon, *Mar. Biol.*, 98, 157–167.
Kvitek, R.G. and Oliver, J.S. 1988. Sea otter foraging habits and effects on prey populations and communities in soft-bottom environments, in Van Blaricom, G.R. and Estes, J.A., Eds., *The Community Ecology of Sea Otters*, Springer-Verlag, New York, 22–47.
Lau, C.J. 1987. Feeding behaviour of the Hawaiian slipper lobster, *Scyllarides squammosus*, with a review of decapod crustacean feeding tactics on molluscan prey, *Bull. Mar. Sci.*, 41, 378–391.

Lauckner, G. 1983. Diseases of mollusca: bivalvia, Kinne, O., Ed., *Diseases of Marine Animals*, Vol. 2, *Ber. Biol. Anst. Helgol.*, pp. 477–984.

Lintas, C. and Seed, R. 1994. Spatial variation in the fauna associated with *Mytilus edulis* on a wave-exposed rocky shore, *J. Moll. Stud.*, 60, 165–174.

Loo, L.-O. and Rosenberg, R. 1989. Bivalve suspension-feeding dynamics and benthic-pelagic coupling in an eutrophicated marine bay, *J. Exp. Mar. Biol. Ecol.*, 130, 253–276.

Loosanoff, V.L. 1956. Two obscure oyster enemies in New England waters, *Science*, 123, 1119–1120.

Lubchenco, J. and Menge, B.A. 1978. Community development and persistence in a low rocky intertidal zone, *Ecol. Monogr.*, 48, 67–94.

MacArthur, R.H. and Wilson, E.O. 1967. *The Theory of Island Biogeography*, Princeton University Press, Princeton, NJ, 203 pp.

Marsh, C.P. 1986. Rocky intertidal community organization" the impact of avian predators on mussel recruitment, *Ecology*, 67, 771–786.

McGrorty, S., Clarke, R.T., Reading, C.J., and Goss-Custard, J.D. 1990. Population dynamics of the mussel *Mytilus edulis*: density changes and regulation of the population in the Exe estuary, Devon, *Mar. Ecol. Prog. Ser.*, 67, 157–169.

McGrorty, S. and Goss-Custard, J.D. 1991. Population dynamics of the mussel *Mytilus edulis*: spatial variations in age-class densities of an intertidal estuarine population along environmental gradients, *Mar. Ecol. Prog. Ser.*, 73, 191–202.

Meire, P. 1993. The impact of bird predation on marine and estuarine bivalve populations: a selective review of patterns and underlying causes, in Dame, R.F., Ed., *Bivalve Filter Feeders in Estuarine and Coastal Ecosystem Processes*, Springer-Verlag, Heidelberg, 197–244.

Meire, P.M. and Ervynck, A. 1986. Are oystercatchers (*Haematopus ostralegus*) selecting the most profitable mussels (*Mytilus edulis*)?, *Anim. Behav.*, 34, 1427–1435.

Menge, B.A. 1992. Community regulation: under what conditions are bottom-up factors important on rocky shores?, *Ecology*, 73, 755–765.

Möller, P. 1986. Physical factors and biological interactions regulating infauna in shallow boreal areas, *Mar. Ecol. Prog. Ser.*, 30, 33–47.

Moore, H.B. and Lopez, N.N. 1969. The ecology of *Chione cancellata*, *Bull. Mar. Sci.*, 19, 131–148.

Moore, H.B. and Lopez, N.N. 1970. A contribution to the ecology of the Lamellibranch *Dosinia elegans*, *Bull. Mar. Sci.*, 20, 980–986.

Murphy, G.I. 1968. Pattern in life history and the environment, *Am. Nat.*, 102, 391–403.

Odum, E.P. 1983. *Basic Ecology*, Saunders, Philadelphia, 613 pp.

Oliver, J.S., Kvitek, R.G., and Slattery, P.N. 1985. Walrus disturbance: scavenging habits and recolonization of the Bering Sea benthos, *J. Exp. Mar. Biol. Ecol.*, 91, 233–246.

Oliver, J.S., Slattery, P.N., O'Connor, E.F. and Lowry, L.F. 1983. Walrus, *Odobenus rosmarus*, feeding in the Bering Sea: a benthic perspective, *Fish. Bull.*, 81, 501–512.

Olla, B.L., Bejda, A.J., and Martin, A.D. 1975. Activity movements, and feeding behavior of the cunner, *Tautogolabrus adspersus*, and comparison of habits with young tautog, *Tautoga onitis*, off Long Island, New York, *Fish. Bull.*, 73, 895–900.

Paine, R.T. 1974. Intertidal community structure: experimental studies on the relationship between a dominant competitor and its principal predator, *Oecologia*, 14, 93–120.

Paine, R.T. 1976. Size-limited predation: an observational and experimental approach with the *Mytilus-Pisaster* interaction, *Ecology*, 57, 858–873.

Perkins, F.O. Infectious diseases of molluscs, in Couch, J.A. and Fournie, J.W., Eds., *Pathobiology of Marine and Estuarine Organisms*, CRC Press, Boca Raton, FL, 255–287.

Peterson, C.H. 1991. Intertidal zonation of marine invertebrates in sand and mud, *Am. Sci.*, 70, 236–249.

Peterson, C.H. and Quammen, M.L. 1982. Siphon nipping: its importance to small fishes and its impact on growth of the bivalve *Protothaca staminea* (Conrad), *J. Exp. Mar. Biol. Ecol.*, 63, 249–268.

Powell, E.N., Cummins, H., Stanton, R.J., and Staff, G. 1984. Estimation of the size of molluscan larval settlement using the death assemblage, *Estuarine Coastal Shelf Sci.*, 18, 367–384.

Powell, E.N., White, M.E., Wilson, E.A., and Ray, S.M. 1987. Small-scale spatial distribution of oysters (*Crassostrea virginica*) on oyster reefs, *Bull. Mar. Sci.*, 41, 835–855.

Rhoads, D.C., Lutz, R.A., Cerrato, R.M., and Revelas, E.C. 1982. Growth and predation activity at deep-sea hydrothermal vents along the Galapagos Rift, *J. Mar. Res.*, 40, 503–516.

Riedman, M.L. and Estes, J.A. 1988. A review of the history, distribution and foraging ecology of sea otters, in Van Blaricom, G.R. and Estes, J.A., Eds., *The Community Ecology of Sea Otters*, Springer-Verlag, New York, 4–21.

Robles, C. 1987. Predator foraging characteristics and prey population structure on a sheltered shore, *Ecology*, 69, 1502–1514.

Rodhouse, P.G. 1978. Energy transformations by the oyster *Ostrea edulis* L. in a temperate estuary, *J. Exp. Mar. Biol. Ecol.*, 34, 1–22.

Rodhouse, P.G., Roden, C.M., Hensey, M.P., and Ryan, T.H. 1984. Resource allocation in *Mytilus edulis* on the shore and in suspended culture, *Mar. Biol.*, 84, 27–34.

Roughgarden, J., Iwasa, Y., and Baxter, C. 1985. Demographic theory for an open marine population with space-limited recruitment, *Ecology*, 66, 54–67.

Sanchez-Salazar, M.E., Griffiths, C.L., and Seed, R. 1987. The interactive roles of predation and tidal elevation in structuring populations of the edible cockle, *Cerastoderma edule*, *Estuarine Coastal Shelf Sci.*, 25, 245–260.

Seed, R. 1969. The ecology of *Mytilus edulis* L. (Lamellibranchiata) on exposed rocky shores. 2. Growth and mortality, *Oecologia*, 3, 317–350.

Seed, R. 1992. Crabs as predators of marine bivalve molluscs, in Morton, B., Ed., *Proc. 1st Int. Conf. Marine Biol. Hong Kong and Southern China*, Hong Kong University Press, Hong Kong.

Seed, R. 1993. Invertebrate predators and their role in structuring coastal and estuarine populations of filter feeding bivalves, in Dame, R., Ed., *Bivalve Filter Feeders in Estuarine and Coastal Ecosystem Processes*, Springer-Verlag, Heidelberg, 149–195.

Seed, R. 1996. Patterns of biodiversity in the macro- invertebrate fauna associated with mussel patches on rocky shores, *J. Mar. Biol. Assoc. U.K.*, 76, 203–210.

Seed, R. and Brown, R.A. 1975. The influence of reproductive cycle, growth, and mortality on population structure in *Modiolus modiolus* (L.), *Cerastoderma edule* (L.) and *Mytilus edulis* L. (Mollusca: Bivalvia), *Proc. 9th Eur. Mar. Biol. Symp.*, pp. 257–274.

Seed, R. and Suchanek, T.H. 1992. Population and community ecology of *Mytilus*, in Gosling, E., Ed., *The Mussel Mytilus: Ecology, Physiology, Genetics and Culture*, Elsevier, New York, 87–169.

Sheldon, R.W. 1965. Fossil communities with multi-modal size-frequency distributions, *Nature*, 206, 1336–1338.

Sims, N.A. and Howard, N. T. 1988. Indigenous tridacnid clam populations and the introduction of *Tridacna derasa* in the Cook Islands, in Copland, J.W. and Lucas, J.S., Eds., *Giant Clams in Asia and the Pacific*, Australian Center for International Agricultural Research, Canberra, 34–40.

Sinderman, C.J. 1990. *Principal Diseases of Marine Fish and Shellfish,* 2nd ed., Vol. 2, *Diseases of Marine Shellfish*, Academic Press, 521 pp.

Skeel, M.E. 1979. Shell-boring worms (Spionidae: Polychaeta) infecting cultivated bivalve molluscs in Australia, *Proc. World Maricult. Soc.*, 10, 529–533.

Smith, J.W. and Merriner, J.V. 1978. Biology and identification of rays in the Chesapeake Bay, No. 20, *VIMS Educ. Ser.*, 22 pp.

Sparks, A.K. 1985. *Synopsis of Invertebrate Pathology Exclusive of Insects*, Elsevier, New York, 521 pp.

Stenzel, H.B. 1971. Oysters, in Moore, R.C., Ed., *Treatise on Invertebrate Paleontology*, Part N, *Mollusca 6, Bivalvia*, Geological Society of America, Lawrence, KS, N954–N1224.

Stuart, V. 1982. Absorbed ration, respiratory costs and resultant scope for growth in the mussel *Aulacomya ater* fed on a diet of kelp detritus of different ages, *Mar. Biol. Lett.*, 3, 289–306.

Suchanek, T.H. 1978. The ecology of *Mytilus edulis* L. in exposed rocky intertidal communities, *J. Exp. Mar. Biol. Ecol.*, 76, 105–120.

Suchanek, T.H. 1986. Mussels and their role in structuring rocky shore communities, in Moore, P.G. and Seed, R., Eds., *The Ecology of Rocky Coasts*, Hodder and Stoughton, Sevenoaks, U.K., 70–96.

Taylor, R.J. 1984. *Predation*, Chapman and Hall, New York, 166 pp.

Trevallion, A. 1971. Studies on *Tellina tenuis* (Da Costa). III. Aspects of general biology and energy flow, *J. Exp. Mar. Biol. Ecol.*, 7, 95–122.

Trevallion, A., Edwards, R.R.C., and Steele, J.H. 1970. Dynamics of a benthic bivalve, in Steele, J.H., Ed., *Marine Food Chains*, University of California Press, 285–295.

Tsuchiya, M. 1983. Mass mortality in a population of the mussel *Mytilus edulis* L. caused by high temperature on rocky shores, *J. Exp. Mar. Biol. Ecol.*, 66, 101–111.

Tsuchiya, M. and Nishihira, M. 1985. Islands of *Mytilus* as habitat for small intertidal animals: effect of island size on community structure, *Mar. Ecol. Prog. Ser.*, 25, 71–81.

Tsuchiya, M. and Nishihira, M. 1986. Islands of *Mytilus* as habitat for small intertidal animals: effect of *Mytilus* age structure on the species composition of the associated fauna and community organization, *Mar. Ecol. Prog. Ser.*, 31, 171–178.

Vahl, O. 1981. Energy transformations by the Iceland scallop, *Chlamys islandica*, from 70°N. II. The population energy budget, *J. Exp. Mar. Biol. Ecol.*, 53, 297–303.

Valiela, I. 1984. *Marine Ecological Processes*, Springer-Verlag, New York, 546 pp.

Vance, R.R. 1973. On reproductive strategies in marine benthic invertebrates, *Am. Nat.*, 107, 339–352.

Van Blaricom, G.R. 1988. Effects of foraging by sea otters on mussel-dominated intertidal communities, in Van Blaricom, G.R. and Estes, J.A., Eds., *The Community Ecology of Sea Otters*, Springer-Verlag, New York, 48–91.

Walford, L.V. 1946. A new graphic method of describing the growth of animals, *Biol. Bull.*, 30, 453–467.

Ward, J.E. and Langdon, C.J. 1986. Effects of the ectoparasitic *Boonea* (=*Odostomia*) *impressa* (Say) (Gastropoda: Pyramidellidae) on growth rate, filtration rate, and valve movements of the host *Crassostrea virginica* (Gmelin), *J. Exp. Mar. Biol. Ecol.*, 99, 163–180.

Webster, J.R. and Medford, R.Z. 1959. Flatworm distribution and associated oyster mortality in Chesapeake Bay, *Proc. Natl. Shellfish. Assoc.*, 50, 85–89.

Wells, H.W. 1961. The fauna of oyster beds, with special reference to the salinity factor, *Ecol. Monogr.*, 31, 241–266.

White, M.E., Powell, E.N., and Ray, S.M. 1988. Effect of parasitism by the Pyramidellid Gastropod *Boonea impressa* on the net productivity of oysters (*Crassostrea virginica*), *Estuarine Coastal Shelf Sci.*, 26, 359–377.

ECOSYSTEM PROCESSES

5 Grazing

INTRODUCTION

In coastal and estuarine ecosystems with dense populations or communities of bivalves, these organisms often represent a major functional component that consumes large quantities of primary producers and strongly couples the benthic and water column environments to each other. In these systems, bivalves graze on phytoplankton in the water column, benthic microflora, resuspended benthic algae, detritus, or all of these. They also can remove one of their potential food competitors, microzooplankton. Thus, the role of the bivalves in moving energy and materials through and within these ecosystems is prominent and the complimentarity between the population and functional approaches to studying ecosystems is often clear.

Early accounts of the food of bivalve filter feeders were the result of gut content analyses of commercially important species. McCrady (1874) reported that the gut contents of *Crassostrea virginica* included diatoms and algal spores, and Lotsy (1893) found similar foods in the guts of oysters, clams, and ribbed mussel. These observations supported the contention of Möbius (1880) that oyster reefs and mussel beds were communities of interacting organisms, with the filter feeders providing the major source of input. Later studies by Morse (1944) noted that the food species might be different in different seasons. The first possible system consequences of filter feeding were speculated upon by Damas (1935) as reported by Lund (1957) and involved self-silting by the cockle (*Cardium=Cerastoderma edule*). It was not until the extensive study of Verwey (1952) on the cockle (*C. edule*) and the mussel (*Mytilus edulis*) in the Dutch Wadden Sea that the potential effects of bivalve filter feeders on an entire ecosystem were addressed.

GRAZING THEORY AS APPLIED TO BIVALVES

Grazing is a system level process that directly connects the herbivores, in this case bivalves, to the food source or primary producers. This process directly reduces standing crop of the primary producers through consumption and results in nutrition to the bivalves, while changing in the community and population structure of the algae. Grazing control on the primary producers is often depicted as a top-down process (see Figure 6, Chapter 1) generated by herbivores higher up the food chain (Sterner 1986). In contrast, nutrient or resource enrichment or limitation and competition are termed bottom-up control of phytoplankton (Tilman 1976; Sommer 1989). Both bottom-up and top-down controls probably take place at the same time (Chapter 1).

As grazing intensity by herbivores on primary producers increases, three potential consequences have been observed (Hilbert et al. 1981). First, there may be a direct linear decrease in the primary producers and primary production; second, there may be a lag period and then a linear decrease; and third, primary production may actually increase and then drop off (Figure 27). The third alternative is known as the "grazing optimization hypothesis" and was first proposed for terrestrial plant–herbivore interactions by McNaughton (1979). In aquatic systems, the herbivore-mediated removal of algal cells may be compensated for or stimulated by (1) increase in light in the water column; (2) the cropping of larger and, thus, older cells in a population, leading to a higher more logarithmic growth rate; (3) biodiversity shifts to faster growing phytoplankton species; (4) higher nutrient recycling rates; and (5) microalgae growth rates may also be stimulated by increased nutrient availability due to reduced storage of nutrients in algae biomass (Prins et al. 1995). The first three of these potential mechanisms is directly the result of grazing, while the others are indirect effects eventually elicited by nutrient enrichment or feedback (see Chapter 6). Because of this mixing of direct and indirect effects, as well as the top-down vs. bottom-up aspects of phytoplankton population regulation, most of the studies concerning these ideas have taken place in controlled experiments, i.e., microcosms and mesocosms, and are more completely explored in Chapter 7.

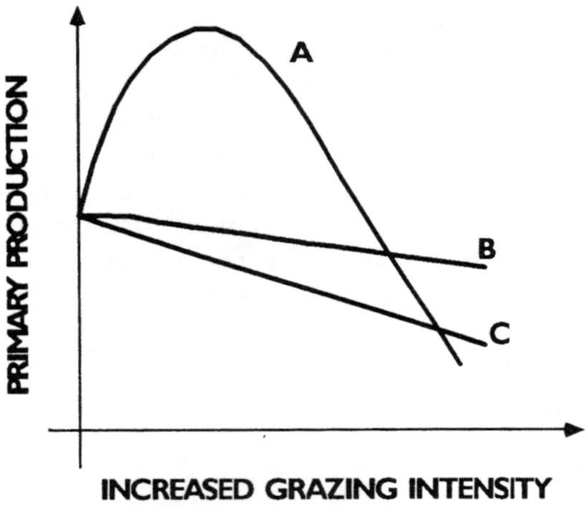

FIGURE 27. Hypothetical influence of grazing intensity on primary production. A = grazing optimization; B = linear decrease; C = time lag followed by linear decrease.

The Officer Conceptual Model for Bivalve Grazing

Officer et al. (1982) used data from South San Francisco Bay to develop a model of bivalve filter feeding as a potential eutrophication control mechanism in shallow coastal bays. These authors recognized that numerous factors had, up to that time, been investigated as controls on the quantity and quality of phytoplankton. Some of these controls were light, temperature, turbidity, self-shading, nutrient concentrations, trace elements, initial seed population, phytoplankton motility and sinking, zooplankton grazing, and hydrodynamic effects. In South Bay, Cloern (1982) was able to discount the preceding factors as major controlling influences and suggested that benthic filter feeders may be the major control mechanism on phytoplankton.

The Officer et al. (1982) model is derived from similar consideration by DiToro et al. (1971), Winter et al. (1975), and Kremer and Nixon (1978). The model assumes that the various constituents interact as in a chemostat or well-mixed system and can be defined as

$$\frac{dc_m}{dt} = \left(k_m - \sum_{n-1}^{N} k_{mn} c_n - k_n \right) c_m + k_h c_{mo}$$

where c_m = concentration of any constituent of the system; N = total number of constituents or dependent variables; c_{mo} = influent constituent concentration to the system; t = time; k_m = net internal production rate of the constituent over natural decay, respiration, or sinking out of the system; k_{mn} = rate coefficients for uptake of c_m by other constituents c_{ni} and k_h = reciprocal to the hydrodynamic residence time (water turnover time). When considering the relationship between the phytoplankton and the filter-feeding benthos, the equations reduce to the Lotka-Volterra predator–prey equations. The equations for this purpose are

$$\frac{dP}{dt} = k_p P - 24 \times 10^{-3} \frac{F}{h} BP$$

$$\frac{dB}{dt} = -k_b B + 24 \times 10^{-6} \alpha \beta_1 \beta_2 FPB$$

where P = phytoplankton population in μg chlorophyll l^{-1}; B = benthic filter feeding population in grams total weight m^{-2}; $k_p = 1/\tau_p$ = plankton growth rate, d^{-1}; $k_b = 1/\tau_b$ = benthic mortality rate, d^{-1}; h = water depth in meters; F = specific filtration rate for a given benthic animal, $l\ h^{-1}$ g total weight of animal^{-1}; α = conversion coefficient, plankton nitrogen: plankton chlorophyll by weight; β_1 = benthic population feeding efficiency, food intake: filtering intake; β_2 = conversion coefficient, total weight of benthic animal: nitrogen weight of animal (Officer et al. 1982).

The first relationship of this pair of equations expresses the continuity relation that the time rate of increase of phytoplankton in a column of water is equal to the rate of production of plankton through photosynthesis in the column minus the rate removal through benthic filter feeding. The second equation expresses the continuity relation that the time rate of increase of the benthic population is equal to the rate of increase of the population through plankton feeding minus the rate decrease of the population through mortality. Values for the various components of the equations are from various sources and are given by Officer et al. (1982). For constant coefficients, the solutions for the pair of equations will oscillate about average values for P and B determined by

$$P = \frac{k_b}{\alpha\beta_1\beta_2 F} \times \frac{10^6}{24}$$

and

$$B = \frac{k_p h}{F} \times \frac{10^3}{24}$$

For the South Bay, $P = 4$ μg chlorophyll l^{-1} and $B = 140$ g m^{-2}. Thus, in a shallow water system, the model predicts that a benthic filter feeding community of a few hundred g m^{-2} will act as a control on the phytoplankton population, and phytoplankton concentrations will be a few μg chl l^{-1}.

The time necessary for the benthic filter feeders to filter the available water or clearance time (τ_f) is given by

$$\tau_f = \frac{h}{FB} \times \frac{10^3}{24}$$

and is approximately 1 d for the South Bay example. The time coefficient for the phytoplankton growth (τ_p) is given by

$$\tau_p = \frac{1}{k_p} = \tau_f$$

Therefore, the two coefficients should be equivalent if the filter feeding benthos are to control the phytoplankton.

From this exercise, Officer et al. (1982) concluded that for the benthic filter feeding community to control the phytoplankton, water depths must be shallow, nutrients must be abundant; there must be no limitation of phytoplankton due to light, temperature, or turbidity; poor hydrodynamic exchange of the water mass (long water mass turnover time) is required; and there must be a dense benthic filter feeding community of small animals.

The Officer et al. (1982) model is a population interaction relationship that includes some environmental constraints. However, because of its use of the functional groups of filter feeders and phytoplankton, it can also be considered a process–functional relationship. The advantage of this simplified modeling approach is that the state of the system can be expressed as a function of a few readily interpretable parameters. Nevertheless, the Officer et al. (1982) model uses the Lotka-Volterra system of equations, and these are difficult to apply to the concept of stability, hard to measure in the field, and lack spatial and temporal variability (Herman 1993). The Officer et al. (1982) conclusion that bivalve control of phytoplankton is only effective in shallow waters depends on the assumption that primary production is independent of water depth and cannot apply to light-limited primary production (Herman 1993). In spite of these problems, the Officer et al. (1982) model does provide a starting point and does raise the issues surrounding the concept of bivalve grazing controling phytoplankton in some systems.

METHODS USED TO ESTIMATE SYSTEM GRAZING

The first estimates of bivalve filter feeder grazing were simply scaled-up from observations made on individuals or small groups of animals in closed systems or jars (see Chapter 3). These methods were limited because they did not take into account the structural effects of aggregations of bivalves nor did they include the role of varying water velocities and food concentrations so typical of natural systems. They did, however, allow the computation of energy and material budgets.

UPSTREAM–DOWNSTREAM OBSERVATIONS

As most dense populations of bivalves occur in flowing water environments, methods developed for these systems eventually were adopted or formed the basis for new approaches to estimating grazing by bivalves. Most of these methods involve some application of the upstream–downstream approach to estimating changes in material concentrations and fluxes.

FREE-FLOW DESIGNS

The upstream–downstream method was first applied to a coral reef by Sargent and Austin (1949) and from an ecosystem perspective by Odum and Odum (1955). In this method, sample stations are selected along a stream of water so that the stream is sampled before and after it passes over a system. It is assumed that the water mass is not influenced in any major way by influxes adjacent to the stream; the flowing water exhibits continuity. In addition to sampling material concentrations, water velocity is measured and cross-sectional water fluxes are computed. The water fluxes are multiplied by the material concentrations to provide material flux estimates both upstream and downstream. If the upstream material fluxes are higher than the downstream fluxes, the system is assumed to have taken up this constituent. If the upstream material fluxes are lower than the downstream fluxes, then the system is thought to be releasing materials into the water. The upstream–downstream method has several advantages: it does not require the removal or disturbance of the benthos; it includes all components of the benthic, water column, and sediment systems; and, environmental conditions are unaffected by the measurements. The most frequent problems with this approach are the lack of water mass continuity; insufficient biomass in the benthic system to produce a measurable response; and, although it is possible to do repeated measurements, there can be no true replication (see Odum et al. 1959 for an extensive discussion).

The upstream–downstream approach was first applied to a mussel bed by Nixon et al. (1971) to estimate the metabolism of the system through differences in oxygen fluxes. Unlike grazing studies, this project had to contend with the exchange of oxygen with the atmosphere. This study also recognized the role of water flow, i.e., current speed, in influencing the metabolism of a benthic system.

Wright et al. (1982) estimated the grazing of *Mytilus edulis* under natural conditions in the Essex Estuary of Massachusetts. Using the classical upstream–downstream design, they assessed changes in bacterial and phytoplankton populations as the flooding tide swept across a mussel bed. Water samples were taken every 5 min by sampling the upstream water and releasing an orange as a "drogue." When the drogue reached the other side of the bed its passage time was noted, and a downstream water sample was taken. Clearance was calculated from these observations using the relationship developed by Jørgensen (1943)

$$r = \frac{1}{t} \cdot \ln\left(C_u / C_d\right) \cdot V$$

where r is clearance, t is the time for the water to pass across the 46.5 m bed, V is the water occupying a 1-m wide path across the bed, C_u is the plankton concentration upstream, and C_d is the concentration downstream. Observations by Wright et al. (1982) clearly show that phytoplankton are removed as they pass across the mussel bed, but bacteria are not. Their field estimates of grazing were 10 times their laboratory estimates derived from individual mussels.

Studies on coastal and estuarine benthic systems (Wildish 1977; Warwick and Uncles 1980; Wildish and Peer 1983) developed the idea that current speed influenced the growth of bivalves through the combined effects of food limitation near the bottom and renewal of food-depleted water by the flow regime. Fréchette et al. (1993) have argued that this paradigm is flawed because it only explains the effects of current speed on growth through correlative means and does not include the mechanistic aspects of current speed and related hydrodynamic characteristics of the environment. They are particularly concerned because the benthic boundary layer, a universal feature of shallow water flows, is not taken into account.

Laboratory Flumes

Laboratory flumes are flow tanks designed to examine processes occurring in steady (constant flow) one-dimensional open-channel flows. In general, the dynamical similarity between laboratory flume and field boundary-layer flows is accomplished by matching four dimensionless flow ratios: the Reynolds number, Froude number, Strouhal number, and Euler number (see Nowell and Jumars 1987 and Fréchette et al. 1993). When dealing with sedentary living bivalve beds interacting with the flowing water both passively (structural roughness) and actively (pumping and filtering water), the use of flumes is tricky and essentially becomes a fluid dynamic engineering exercise. With bivalve beds, large flumes are required to provide (1) one-dimensional flow, (2) fully developed flow (boundary layer reaches the surface), (3) fully developed turbulent flow, and (4) that the effects of the bed on phytoplankton concentrations and the effects of this food supply on bivalve feeding and growth rates requires bivalves covering a major portion of the flume in the along-channel direction. As a rule of thumb, Fréchette et al. (1993) recommend a flume at least 50 cm wide by 10 cm deep by about 10 m long.

The depletion of suspended food above dense beds of bivalves has been closely examined through the use of such laboratory flumes (Wildish and Kristmanson 1984; Fréchette et al. 1989, 1993; Monismith et al. 1990; Butman et al. 1994). These studies clearly showed that the benthic boundary layer is important because seston depletion decreases as a function of flow decreasing within the benthic boundary layer. It is the combination of vertical mixing and horizontal advection that determines food supply to the bivalve bed. Rather than food concentration controlling growth, several studies (Grizzle and Lutz 1989; Grizzle and Morin 1989; and Muschenheim and Newell 1992) have argued that the controlling factor is food flux. The issues of food concentration or food flux regulating bivalve growth are as yet unresolved.

Because of the intensity of sampling necessary to determine vertical and horizontal gradients across bivalve beds in the field, Fréchette et al. (1989) developed a steady, two-dimensional, numerical, finite-difference model to evaluate the magnitude of these gradients of food over an intertidal mussel bed. This model represents the balance between horizontal advection and vertical diffusion of phytoplankton over a mussel bed, i.e., the spatial variation of phytoplankton concentration due to mussel grazing and boundary-layer flow. The model assumes that time variations are small, as are sources and sinks internal to the water column (primary production and zooplankton grazing). The simplified model takes the form:

$$u\frac{\partial F}{\partial x}+\frac{\partial}{\partial z}\left(-K\frac{\partial F}{\partial z}\right)=0$$

The first component is the horizontal advection of phytoplankton, where u is the horizontal velocity, F is the phytoplankton concentration, and x is the downstream distance along the mussel bed; and the second term is the vertical diffusion where K is vertical eddy diffusivity and is assumed to be equal to the eddy viscosity, and z is the height above the bottom.

In the Frechette et al. (1989) study, consumption of phytoplankton by the mussel bed was estimated from the uptake of fluorescence by mussels in a flow-through chamber and scaled-up to total bed consumption by multiplying the weight-specific uptake of fluorescence times the size class biomass of mussels in the bed. Fluorescence consumption was found to be correlated to fluorescence F as computed in the horizontal advection and vertical diffusion model.

Using this combination of flow-through filtration chamber and gradient models and under typical flow and roughness conditions, the Fréchette model predicts a significant decrease in phytoplankton concentration in the water over the mussel bed. Vertical gradients are stronger than horizontal gradients, and they are steepest close to the bed. Uptake of phytoplankton by the mussel bed results in a gradual decrease in phytoplankton concentrations downstream. Mussel bed consumption rate at the center of the bed increases almost linearly with increasing velocity, and it also increases with increasing bed roughness. Increasing height of mussel intake above the bottom also increases consumption. This attribute supports the formation of mussel hummocks to increase food consumption, presumably with enhanced growth.

The Fréchette et al. (1989) model has been supported by several very graphical experiments using a 17-m laboratory flume (Butman et al. 1994). Not only did this upstream–downstream experimental design allow a multitude of vertical and horizontal water sampling points for phytoplankton (fluorescence) concentrations, but velocity was extensively measured with a two-axis forward-scatter laser Doppler velocimeter, also known as a two-component laser Doppler anemometer (LDA). Their observations show significant enhancement of turbulent stress due to mussel bed roughness and provide further evidence of horizontal and vertical depletion of the food supply to the mussel bed. Fréchette et al. (1993) have argued that pumping jets from mussels do not influence the boundary flow to any great extent.

Epifaunal bivalves are not the only animals to interact with the benthic boundary layer. Infaunal siphonate bivalves may also influence the near-bottom flow (Monismith et al. 1990; Koseff et al. 1993), both by their physical structure and biotic pumping. Using a laboratory flume with model clams, an LDA and laser-induced fluorescence (LIF) system, this group has investigated the role of pumping jets and siphon behavior of infaunal clams on near-bottom flow. It appears that a significant proportion of excurrent water could be refiltered, depending on boundary layer velocity, excurrent water velocity, and siphon height, orientation, and size. They also showed that the interaction of bivalve siphon currents with the overlying turbulent boundary layer flow results in the formation of concentration boundary layers over dense populations of clams (O'Riordan et al. 1993). Clam feeding within a bed appears to benefit by increasing siphon height and decreasing pumping speed as velocity increases.

Working with this group, Cole et al. (1992) developed a recirculating laboratory flume that was used to investigate the filtration rates of the infaunal bivalve *Potamocorbula amurensis*. Following the design criteria of Nowell and Jumars (1987), this flume allows continual recirculation of water over a test section where grazing bivalves and their particle capture are localized. The main channel is 350 cm long and 38 cm wide, with the test section 250 cm from the head of the flume. A portion of the flume water and suspended algae are continuously withdrawn from the end of the flume, passed through a fluorometer and then back into the flume. Vertical profiles of the flow velocity confirm the presence of a well-formed boundary layer over the test bed. As suggested by Monismith et al. (1990), friction velocity and roughness height were measured as the critical parameters necessary to match natural flow environments and found to be within those observed in estuaries. Using the cultured phytoplankter *Chroomonas salinay* and conducting measurements over a free velocity range of 9 to 25 cm s^{-1}, the filtration rate of the assemblage of different size clams was determined. Clam filtration rates appeared to be sensitive to water velocity and support the observations of Wildish and Miyares (1990) that increasing water velocities inhibit filtration and that growth of clams is generally inhibited at flows above 15 cm s^{-1} (Wildish and Kristmanson 1993).

Field Flumes

Field flumes offer a straightforward method to directly measure grazing as well as other material fluxes by dense beds of bivalves. These structures offer several advantages (Fréchette et al. 1993): (1) the benthic community and system are not disturbed; (2) fluxes are computed for a large area of the system, not just an individual or a few individuals; (3) the effects of changing flow can be directly observed; (4) general environmental conditions are not constrained; and (5) flow is simplified or channelized by the walls of the flume. These authors further caution that flumes should (1) not interfere with the natural benthic boundary layer; (2) be aligned with the current flow; and (3) be of a size and configuration so as to avoid wall or edge effects.

The only successful field flume I am aware of is the Sylt flume (Asmus et al. 1990). This flume was constructed across an intertidal bed of mussels (*Mytilus edulis*) in the Eastern Wadden Sea near the island of Sylt, Germany. The flume was oriented along the axis of the main tidal currents, and water entering and leaving the flume had little direct contact with other mussels. A wooden support structure was anchored upon the mussel bed. The flume was 20 m long, 2 m wide, and 2 m high. Platforms were constructed above high water at each end of the flume to hold scientists and instruments. The wooden superstructure of the flume remained permanently in place during the summer sampling period. During sampling, polyethylene sheeting was attached to the wooden frame along each side to form a channel. To prevent leakage underneath the wall of the flume, the plastic sheeting was sealed to the bottom with heavy iron chain. Thus, a 40 m^2 area of mussel bed with over 140,000 mussels was channelized, and lateral currents were prevented. The tidal range in this location was 1.8 m and the mussels were submerged 8 to 10 h on each tide.

Water velocity was measured using plastic drift bags filled with freshwater or seawater. The transient time of the drift bags through the flume was an estimate of water flow. This drift bag method of water flow through the flume was calibrated against direct current estimates using an induction flow meter, and the two sets of measurements were found to be in close correspondence (Asmus et al. 1990). In spite of its structural size, this flume meets the requirements laid out by Fréchette et al. (1993) for incorporating the benthic boundary layer flows.

TUNNELS

Tunnels can be described as flumes with a top and have been utilized in high tidal range and water flux environments. The original tunnel for bivalve research was developed by Dame et al. (1984) to investigate the fluxes of materials between the water column and intertidal oyster reefs. The tunnel was the culmination of a series of efforts to measure these material fluxes in a high tidal flow environment at North Inlet, South Carolina. This group's initial efforts focused on a free-flow upstream–downstream design within a tidal creek that had dense populations of oysters. That effort failed to show significant differences in material fluxes across the oyster beds in question, probably due to a lack of continuity in the water mass from upstream to downstream. Next a flume very similar in size and configuration to the Sylt flume was constructed across an oyster reef in the same area as before. Again there was difficulty in determining significant differences in the material fluxes across the reef and it was thought that the volume of water passing through the flume was sufficiently large to dilute the materials of interest. It was at this point that these investigators became aware of Rogers' (1979) study of a coral reef using a tunnel and following in upstream–downstream approach pioneered by Odum and Odum (1955) and Nixon et al. (1971). The Rogers (1979) tunnel was used to measure the productivity of a coral reef, but it did offer a way to ensure continuity of water flow and reduce the influences of dilution.

With the preceding knowledge, the North Inlet group constructed a portable 10 m long plastic tunnel that was installed over an oyster reef (Dame et al. 1984). The tunnel (benthic ecosystem tunnel — BEST) was made of eight formed Plexiglass plates, and each section was joined in the field with neoprene strips. On the first tunnels, neoprene was also used to seal the tunnel to the bottom. Later versions of the tunnel were adapted to rest on permanently installed steel rails (Prins and Smaal 1994). BEST was 0.8 m wide and had a cross-sectional area of 0.225 m^2. It covered 7.9 m^2 of bottom and had a nonflow volume of 1.74 m^3.

As it was impossible to take continuous multiple water velocity measurements within the tunnel, Kjerfve and Dame (unpublished; also see Prins 1996) conducted a detailed study of the water flow dynamics within the tunnel while it was deployed over an oyster reef and mussel beds. Using an induction flow meter on a mobile mount, flow across the entire cross section and length of the tunnel was observed, and it was determined that total cross-sectional water flux was strongly related to water flow measured at a single point in the center of the tunnel. Later dye studies on a similar tunnel in the Netherlands showed that vertical mixing (turbulent flow) was observed at all water velocities >1 cm s^{-1} (Binsbergen, unpublished; Prins 1996).

Material concentrations were determined from water samples taken simultaneously from 0.5 m within each end of the tunnel and, as there was continuity of water flow through BEST, the mass fluxes across each end of the tunnel were calculated by multiplying the water flux by the incoming and outgoing material concentrations. The net flux of a particular constituent was the difference between incoming and outgoing mass fluxes. If the net flux had a positive value, material uptake occurred as the water passed through the tunnel and over the bivalves. If the net flux was negative, material had been released. Since the incoming water had not been within the tunnel, it was considered to be a control or untreated sample, while the outgoing water sample was considered to be an experimental or treated sample. The input and output water samples were compared statistically using paired-t test to examine for significant differences. Later studies used a tunnel with no filter feeders present as a parallel control (Asmus et al. 1993). It was assumed that the tunnel had no effect on the biogeochemistry of the sampling site.

In the only comparison of material flux techniques to date, the Dutch tunnel group and the Sylt flume group compared these two measurement systems simultaneously (Asmus et al. 1992). In general, the tunnel measurements were indicative of the minimum material flux, were more flexible in installation including time requirements, and could be used in deep waters. To avoid trapping and sedimentation effects, tunnels must be adapted to the specific current regime. Flumes can only be used in shallow waters with stable current regimes. These investigators concluded that the inferences concerning the ecological role of mussel beds is not likely to be different, whether tunnels or flumes are used.

ECOSYSTEM GRAZING ESTIMATES

The following discussion will address those systems in which grazing by bivalves has been estimated. The available descriptions of the systems are not uniform, and for this reason there is some inconsistency in the comparisons. The systems are discussed in alphabetical order.

Askö, Sweden

The Askö research area comprises a relatively unpolluted archipelago of the Swedish Baltic coast and is approximately 160 km^2. Because of the comparatively low salinities 6 to 7‰ above the halocline, the ecosystem is relatively simple with few species. *Mytilus edulis* makes up about 80% of the animal biomass within the ecosystem (Kautsky and Wallentinus 1980) and most of these bivalves are subtidal. Using field data and plastic bag enclosures (Kautsky and Wallentinus 1980; Kautsky 1981; Kautsky and Evans 1987), this group argued that the high grazing capacity of the mussels will have a stabilizing effect on the ecosystem. This grazing capacity is clearly demonstrated at the start of the phytoplankton spring bloom, when mussel gonad build-up is immediately initiated, resulting in as much as 8% of the bloom being stored as gonad. Thus, the mussels respond to the phytoplankton bloom immediately, while zooplankton must go through a complete life cycle before they can have a significant effect on the bloom. In spite of the dominance of mussels in this system, they can only clear the water volume in question in about 100 days, while the water residence time is in the order of 3 years. Thus, these findings do not seem to support a strong influence of mussels on phytoplankton in this ecosystem.

Bay of Brest, France

The Bay of Brest is located on the northwest Atlantic coast of France. It is an area of about 102 km^2 connected to the sea via a narrow channel. The bay is dominated by muddy banks, slopes, and channels. Phytoplankton primary production is high (220 gC m^{-2} $year^{-1}$), but phytoplankton biomass is usually comparatively low (1 to 1.5 mg m^{-2} chlorophyll *a* [Chl *a*]). Hily (1991) investigated the proposition that the bay either exports phytoplankton to the adjacent sea or the phytoplankton is consumed by grazers within the system. Zooplankton are discounted as major grazers within the system and benthic grazers including pectinid clams (*Chlamys*), barnacles, and sponges are implicated as the major phytoplankton consumers. From survey data and filtration rates in the literature, Hily (1991) concluded that the benthic community was capable of filtering the water in the bay with a time constant shorter than the hydrodynamic residence time and comparable with the time constant for phytoplankton growth. He further notes that suspension feeding epifauna are more important than suspension feeding infauna in the filtration of the bay. The impact of the filter feeders progressively decreases from shallow to deeper water environments, possibly because water volume increases with depth.

Chesapeake Bay, U.S.A.

Chesapeake Bay is the largest estuarine water body in the U.S. (11,500 km^2) and is located on the central Atlantic coast. With an average depth of 7 m, a water volume of 27.3×10^9 m^3, and a residence time of 22 days, the bay is extensively used for fisheries, recreation, and waste disposal.

Historically, the bay was the primary commercial source of the American oyster, *Crassostrea virginica*, but over the past century, and particularly in the time period from 1970 to present, declining water quality, disease, and overharvesting have led to the precipitous reduction of this fishery (Newell 1988; Ulanowicz and Tuttle 1992). A recent study by Rothschild et al. (1994) concluded that the loss of the oyster fishery was mainly attributable to overharvesting and a destruction of oyster reef habitat as a result of harvesting methods, and that disease and declining water quality only contributed to the decline.

In an interesting comparison of filtration capacity, Newell (1988) calculated the time it would take for oysters to filter or clear the entire volume of the bay in the pre-1870 era and in 1988. Not surprisingly, there was a 100-fold increase in the clearance time from 1870 to 1988.

With the decline of the oysters, Newell (1988) argues that the major filter feeding competitor of the zooplankton was removed, and these pelagic suspension feeders increased. This increase also allowed the development of the pelagic food web with its now common predators, ctenophores and jellyfish. He further contends that the historically large populations of oysters not only cropped phytoplankton, but they were probably a major agent in controlling the present turbidity and anoxia problems. Newell (1988) points out that without the year round filtration of the benthic oysters and with the low density of zooplankton in the spring, the spring phytoplankton bloom in the bay is largely ungrazed. The ungrazed phytoplankton eventually sinks to the bottom and provides the organic fuel for the anoxic conditions in the bottom waters of the bay (Kemp and Boynton 1984).

Concurrent with the decline of the oyster population, numerous bivalve species have invaded the bay, e.g., *Rangia cuneata* and *Corbicula fluminea* (Newell, personal communication). In an interesting modeling exercise, Gerritsen et al. (1994) have shown that these invaders and other filter feeding bivalves including *Macoma* spp., *Mulinia lateralis*, *Mya arenaria*, and *Tagelus plebeius* are capable of removing up to 50% of the annual phytoplankton primary production in the upper Chesapeake Bay. They contend that rehabilitating the oyster populations may not change the water quality in the bay, because these bivalves will only be capable of filtering the water in the shallow zones. From this review, it is clear that the role of bivalves in modern Chesapeake Bay is still being sorted out.

Delaware Bay, U.S.A.

Delaware Bay is located in the mid-Atlantic region of the U.S., with a surface area of 1942 km^2 and a water volume of 19.4×10^9 m^3. The bay has large areas of open water fringed with flats and marshes. Both planktonic and benthic biomass are considered low relative to other estuarine systems (Biggs and Howell 1984). The major filter feeders in the bay are zooplankton dominated by *Acartia tonsa, Temora longicornus, Centrophages hatamus*, and *Pseudocalanus minutus*. Biggs and Howell (1984) estimated that these zooplankton can filter the entire bay volume in 77 days or about 3 to 5 times a year. The dominant bivalve filter feeder is *Crassostrea virginica*, but, as in Chesapeake Bay, populations of this organism have been severely depleted over the past century due to overfishing and disease (Maurer et al. 1971). Current populations of bivalves are so low that it would take 3.5 years for them to filter the water mass in Delaware Bay or about 10 times longer than the physical residence time of water in the bay. Consequently, in this bay, bivalves probably do not have much influence on the phytoplankton populations, while zooplankton do.

Killary Harbour, Ireland

Killary Harbour is a fjord type inlet on the west coast of Ireland, about 7.7 km^2 in area with a volume of 102×10^6 m^3 (Rodhouse and Roden 1987). Mussels are intensively cultivated from suspended rafts, and these investigators utilized energy budgets to determine the amount of food necessary to support these animals. Indirect evidence from these authors implies that there must be an influx of phytoplankton into the fjord, and this input can support additional mussel culture. Using filtration rates and biomass estimates given by Rodhouse and Roden (1987), Smaal and Prins (1993) calculated that it would take about 140 days for the mussels to clear the water volume in this system. As the system is fjord-like, the residence time of water in this system is probably weeks to months in magnitude, and inputs from the adjacent sea may not be as important as first thought. As with Ria de Arosa, Killary Harbour is a naturally plankton-dominated system that is being forced into a different trophic structure by the intensive culture of normally benthic bivalve filter feeders.

Laholm Bay, Sweden

Even though bivalve filter feeders may be the dominant benthic organisms, they may not always control phytoplankton biomass. In Laholm Bay on the Kattegat of Sweden, Loo and Rosenberg (1989) investigated the benthic-pelagic coupling between the bivalves *Cerastoderma edule* and *Mya arenaria* and the plankton. The bay is 300 km^2 and about 20% is 0 to 10 m deep. The deeper waters of the bay are anoxic. Using field data on plankton and benthic bivalve populations and laboratory and literature data on feeding, growth, respiration, and excretion, these investigators developed energy budgets for the two bivalve species. Although the bivalves can theoretically filter all of the water volume in the bay down to 10 m in 3 days, they actually only filter about half of that. Of the phytoplankton consumed in the bay, the bivalves remove 97% while the zooplankton account for 3%. A majority of the phytoplankton is exported from the bay. These authors argue that the effectiveness of the bivalves in controlling phytoplankton biomass is constrained by mass mortality in the bivalve populations due to eutrophication effects and poor vertical mixing of the water column. As discussed later, the shifting of the system from bivalve dominance to some other state as a result the loss of the bivalve component is poorly understood at the ecosystem level.

Lowes Cove, Maine, US.A.

In a similar vein, Carlson et al. (1984) studied the influence of a mudflat, Lowes Cove, adjacent to the Gulf of Maine on phytoplankton in the overlying water column. Using observed seasonal and tidal chlorophyll concentrations both within the cove and in the external channel, these investigators found that the cove was a sink for chlorophyll (e.g., phytoplankton). As the chlorophyll did not accumulate in the sediments and observed zooplankton densities were low, settlement and zooplankton grazing were discounted as significant influences on the phytoplankton. Because there was a persistent seasonal pattern in the removal of the phytoplankton on the waters flooding Lowes Cove and this pattern was in phase with the growth of the resident filter feeding bivalves, they suggested that it was the benthic filter feeders living in the cove that were consuming the phytoplankton. These bivalves, *Mytilus edulis* and *Mya arenaria*, are sufficiently numerous to filter over 50% of the Lowes Cove tidal volume, and Carlson et al. (1984) stress that their data not only implicate these animals as major agents in the removal of phytoplankton, but also in the removal of zooplankton.

Bay of Marennes-Oléron, France

The Bay of Marennes-Oléron is a highly variable shallow water body along the western Atlantic coast of France. The bay is 135.7 km^2 with a water volume of 675 $\times 10^6$ m^3. Marennes-Oléron is the major European area for oyster (*Crassostrea gigas*) cultivation. Mussels (*Mytilus edulis*) are also cultivated on a lesser scale. From 1960 to 1969, growth and production of bivalves declined rapidly, due to overexploitation of the food resources, and culminated in an epidemic of iridovirus that wiped out the *C. angulata* population. Current populations of *C. gigas* were introduced, but overstocking again occurred (Héral et al. 1988; Bacher 1989; Héral 1993). Calculations by Smaal and Prins (1993) show that the bivalves can filter the water volume within the bay in about 2.7 days, while the residence time of the water is about 7.1 days. This system is unique in that it is one of the few documented occurrences of overstocking with bivalve filter feeders on two different occasions with different species.

Narragansett Bay, Rhode Island, U.S.A.

Narragansett Bay is located along the coast of Rhode Island, with an area of about 265 km^2. This bay has been the focus of ecological studies for many years and is one of the first systems to be simulated with a comprehensive ecosystem oriented computer model (Kremer and Nixon 1978). Although the bay is dominated by plankton processes, large widespread populations of the hard clam, *Mercenaria mercenaria*, dominate the benthic communities. In the Kremer and Nixon (1978) formulations, clams consume via filtration 29% of the net primary production in the bay and zooplankton consume 79%. Using these authors' data, the water volume of the bay is approximately 2.4×10^9 m^3 with a tidal prism of 310×10^6 m^3. The residence time of the bay water volume is 27 days. The clam populations in the bay can potentially filter 15% of each tidal prism or the entire bay volume in 25 days.

Thus, Narragansett Bay offers an example of a coastal system where bivalve grazing probably does not dominate ecosystem processes, but appears to be balanced with planktonic grazing processes.

North Inlet, South Carolina, U.S.A.

North Inlet is located on the northeast coast of South Carolina and is dominated by extensive salt marshes, tidal creeks, and dense populations of intertidal oysters, *Crassostrea virginica*. The system is about 32 km^2 in area and has a volume of 22 $\times 10^9$ m^3. It has a major connection to the Atlantic ocean and several minor connections to an adjacent estuary to the south. An ecosystem scale study concluded that filtration by oysters strongly coupled the benthic and water column subsystems (Dame et al. 1980, 1986). In this system, concentrations of phytoplankton measured by Chl *a*, phytoplankton plasma volume, and total biomass as estimated by ATP consistently showed high values at high tide and rapidly declining values as the tide ebbs (Chrzanowski et al. 1979). Concurrent analysis of suspended sediments indicated an organic fraction that behaved similarly to ATP and Chl *a* except on spring tides. These observations suggest that the North Inlet ecosystem is importing phytoplankton from the coastal ocean and supports similar findings by Moll (1977) on the Flax Pond ecosystem on Long Island, New York.

Both settling and zooplankton consumption were discounted as possible explanations of North Inlet ecosystem consumption of phytoplankton (Dame et al. 1980). Rough estimates of total oyster filtration in North Inlet showed that this component of the ecosystem was capable of pumping 1.5×10^7 m^3 per tide of a total tidal prism of 2.2×10^7 m^3. Thus, the dense oyster reefs in North Inlet are capable of influencing the suspended particulate environment on a relatively short time scale, the time of a single tidal cycle.

Oosterschelde Estuary, The Netherlands

The Oosterschelde (Eastern Scheldt) estuary is located on the southwest coast of the Netherlands and is part of the Dutch delta system where the Rhine River exits to the North Sea. The water volume of the estuary is about 2.7×10^9 m^3 and has a residence time of 20 to 135 days depending on physical conditions and distance from the North Sea (i.e., faster turnover closer to the North Sea). The estuary is an international wetland, including sand- and mud-flats, cultivating large populations of mussels and cockles, feeding large numbers of migratory waterfowl, and further comprising deep channels, artificial rocky shores, and salt marshes. The system is unique because it is the site of one of the most significant engineering projects of the 20th century, the storm surge barrier with its associated hydrotechnical works. As a result of the barrier-induced changes in the Eastern Scheldt, this system has been extensively studied and is synthesized by Nienhuis and Smaal (1994).

The dominant food chain in the Eastern Scheldt consists of phytoplankton, bivalve filter feeders, wading birds, and human harvesters. The phytoplankton and filter feeder components are similar both before and after the structural changes to the estuary. Smaal et al. (1986) first estimated the effects of bivalve grazing on this system. Later, using the BEST tunnel technique, Dame et al. (1991a) estimated that the mussels within the Eastern Scheldt can clear the water column of phytoplankton in about 5 days or in considerably less time than the water mass turns over. This high system clearance rate implies the control of the phytoplankton populations by the dense populations of bivalve filter feeders. In addition, Herman and Scholten (1990) have shown that this effective top-down control by grazing can keep phytoplankton biomass low and stabilizes the ecosystem as long as the primary producers do not escape filter feeder control by shifts to species that are not or are less efficiently filtered, like macrophytes or very small cells (Riemann et al. 1988).

Ria De Arosa, Spain

The Ria de Arosa is an ocean-dominated bay along Spain's northwest Atlantic coast. The Arosa is the largest of several bays (230 km^2) and supports the intensive raft culture of *Mytilus edulis*. Compared to many of the other systems reviewed here, the Arosa is unique because of the intermittent upwelling of nutrient-rich deep Atlantic Ocean water into the bay. This enrichment supports high water column primary production that in turn supports the culture of mussels (Tenore et al. 1982). Mainly due to the huge volume of organic sediments produced beneath the mussel rafts, Rias with intensive mussel culture have quite different benthic environments than those without aquaculture. Smaal and Prins (1993) combined the data of Tenore et al. (1982), Navarro et al. (1991), and Perez Camacho et al. (1991) to generate estimates of the filtration capacity of *Mytilus* in the Ria de Arosa. Their calculations showed that the mussels could filter or clear the Arosa in 12.4 days or in about half the time the water remained in the system.

San Francisco Bay, California, U.S.A.

San Francisco Bay is a large estuarine system on the central coast of California. The Bay is divided into two hydrologically distinct arms: the northern bay, a typical riverine estuary that receives fresh water from the large Sacramento–San Joaquin river system, and the southern bay, a lagoon-type estuary where fresh water is limited to local run-off sewage effluent and the influx of northern bay water during very high frequency freshwater inflows during winter (Thompson and Nichols 1988). The Bay has a surface area of approximately 1240 km^2 with an average depth of 2 m (including mud flats) or 6 m (excluding mud flats) and a residence time of 11.1 days to several months in summer.

San Francisco Bay has experienced numerous changes in its bivalve populations and biodiversity since the Europeans arrived in the mid-1800s. Initially, the Bay was dominated by populations of *Ostrea lurida*, *Mytilus edulis*, and *Macoma nasuta*. The Europeans did not like the taste of *O. lurida* and imported *Crassostrea virginica* from the East Coast of the U.S. Other bivalves were incidentally imported, including *Mya arenaria*, *Geukensia demissa*, *Tapes philippanarum*, and *Musculista senhousia*. Gradually, the *O. lurida* populations declined, probably because increased suspended sediment loads reduced settlement success. Finally, in 1986, the Asian clam, *Potamocorbula amurensis* invaded and, because it is tolerant of all salinities, depths and substrate types, it has achieved dominance everywhere in the estuary (Nichols and Pamatmat 1988).

Cloern (1982) reported as part of a continuing series of studies on the regulation of the phytoplankton biomass in South San Francisco Bay. Using a one-dimensional dispersion model (1-D) with spatial (transects) and temporal (seasonal) observations of many variables including Chl *a* concentration, the net rate of phytoplankton population growth was estimated. The model estimates showed little relationship to the observed phytoplankton population dynamics in the South Bay. Cloern (1982) evaluated his model and concluded that none of the assumptions inherent in the calculation of the specific growth rate of phytoplankton were grossly violated. He also found that zooplankton grazing and physical transport from the system were of insufficient magnitude to explain the discrepancies. Cloern (1982) noted that benthic invertebrates are abundant in South San Francisco Bay and that the suspension feeding bivalves, *Musculus senhousia, Tapes japoonica*, and *Gemma gemma*, were common. He estimated that these three bivalves could potentially filter 7.2 to 10.8 m^3 m^{-2} d^{-1} in a portion of the South Bay averaging only 6 m deep and, thus, filter a volume equivalent to 1.2 to 1.8 times the South Bay volume each day. Cloern (1982) also noted that as the South Bay occasionally stratifies, the benthos are effectively separated from the surface waters and removed from grazing control. Thus, stratification allows potential bloom conditions for the phytoplankton.

Later studies by Doering and Oviatt (1986) criticized Cloern's (1982) estimates of the filtration capabilities of the bivalves in South Bay. They noted that in their controlled studies Cloern's filtration rate method yielded estimates an order of magnitude above actual observations. Doering and Oviatt (1986) attribute these differences to the use of filtration rates determined in the laboratory using monocultures of phytoplankton. In spite of these differences, they agree that bivalve filter feeders can represent important control mechanisms on phytoplankton populations.

In northern San Francisco Bay, Nichols (1985) has also invoked the potential control of phytoplankton by benthic filter feeders. Northern San Francisco Bay is strongly influenced by river flow from the Sacramento and San Joaquin rivers and is an area of high chlorophyll concentrations and water column primary production. It supports an important planktonic food web involving mysid shrimp and striped bass larvae (Cloern et al. 1983). During a period of extended drought (1976 to 1977) phytoplankton biomass decreased dramatically. Initially, nutrient limitation, light availability, and zooplankton grazing were examined and rejected as potential explanations and greatly reduced river flow was favored. An examination of benthic survey data from 1972 through the drought period showed that the clam, *Macoma balthica*, increased during the drought period and its increase was accompanied by the temporary invasion of this area by a community dominated by the soft-shell clam, *Mya arenaria*. Using bivalve filtration rates determined by Møhlenberg and Riisgård (1979) and bivalve population data from his own observations, Nichols (1985) concluded that the *Mya* population could filter the water column in 0.5 to 2 days, depending on the season and specific location. Nichols (1985) further observed that during any prolonged period of low river flow into north San Francisco Bay, an important natural change, the system will shift from a pelagic to a benthic food web in which energy passes directly from the primary producers in the water column to the benthos.

Continuing studies in northern San Francisco Bay (Alpine and Cloern 1992 and Nichols et al. 1990) support Nichols' (1985) contentions that grazing by filter feeding bivalves can control phytoplankton concentrations, but on a longer time scale. In this case, another bivalve filter feeder, *Potamocorbula amurensis*, invaded the North Bay area beginning in 1987. An examination of the benthic grazer populations from 1974 to 1990 showed that when the density of these animals was low, there were summer phytoplankton biomass maxima, but in years when there were high densities of benthic bivalve grazers, there was no summer phytoplankton biomass peak. This clam has also been implicated in the decline in zooplankton populations in the same area of the bay which is attributed to their predation on the nauplii of mainly copepods (Kimmerer et al. 1994). Alpine and Cloern (1992) examined other possible explanations for the decline in phytoplankton biomass in the North Bay, including resource limitation, toxic contaminants, and reduced phytoplankton growth rates, and concluded that none of these were operational. They believe that their observations support the control of phytoplankton production by the joint effects of direct physical processes (e.g., river-driven transport) and trophic interactions (episodes of enhanced grazing pressure by immigrant populations of benthic suspension feeders). They further concluded that in this estuary top-down and bottom-up forces work in concert to regulate phytoplankton dynamics at seasonal and inter-annual time scales.

Eastern Wadden Sea, Sylt, Germany

The island of Sylt lies in the eastern part of the North Sea and provides protection from ocean waves and winds to the Eastern Wadden Sea. On the northern lee side of Sylt lies the bay of Königshafen, which is a shallow intertidal flat area interlaced with tidal channels. Extensive intertidal mussel beds of *Mytilus edulis* occur at the low tide mark. The biology and ecology of these flats have been extensively studied and are summarized by Reise (1985). More recent work by Asmus (1987), Asmus et al. (1990), and Asmus and Asmus (1991) focused on the role of the filter feeders in interacting with the water column in the Königshafen ecosystem. Using a 20 m long flume, the Asmus group was able to determine filtration rates for mussel beds as a whole. Phytoplankton biomass was reduced 37% between the inflow and outflow of the flume (Asmus and Asmus 1991). The entire range of phytoplankton cells from 4 µm to several hundred µm was reduced by the filtering action of the bed. There was a significant positive correlation between both concentration and uptake of phytoplankton by the mussel bed. From their data (Asmus et al. 1990), the mussels appear to be able to clear the water volume of the system in about 2.1 days, while the water volume itself turns over every tide (this system is mainly intertidal). Thus, the mussel beds filter somewhere between 25% and 50% of the water in the ecosystem on each tide.

Western Wadden Sea, The Netherlands

The Western Wadden Sea is located in northern Holland and is separated from the North Sea by the islands of Texel, Vlieland, and Terschelling. Water exchanges between the North Sea and the Wadden Sea through inlets between North Holland and Texel, Texel and Vlieland, and Vlieland and Terschling. Freshwater enters the sea via runoff from the adjacent mainland. The Western Wadden Sea has a volume of about 4×10^9 m^3 with a residence time of 5 to 15 days depending on climatic conditions. There are extensive intertidal flats interlaced with creeks. Beds of bivalve filter feeders, mussels and cockles, are abundant.

Verwey's (1952) study was based on the premise that the Wadden Sea was inhabited by large numbers of animals and that as the tidal flats within this system were changing rapidly, it was important to know what role these animals played in the formation of or the change in the tidal flats. With a truly ecosystem approach, Verwey (1952) described the ecology of the cockles and mussels in the Dutch Wadden Sea, including their preferred habitat, the role of the physical environment in controlling their feeding, the potential foods available in the water column, and the materials and algae within their guts. He then scaled up from the filtration rates and shell formation rates of individual bivalves to estimate the total amount of material potentially removed from the water column, and available to restructure the tidal flats within the Wadden Sea. This amount, 250×10^6 kg (dry wt) per year, is far greater than that available within the Wadden Sea and suggests that material must be imported from the North Sea. Later observations by Jonge and Postma (1974) and Van der Eijk (1979) showed that chlorophyll and primary production were remarkably low in areas of extensive mussel culture and that this apparent result of bivalve grazing was greatest during the summer when mussel growth was highest. It is interesting to point out that while the bivalves in Verwey's (1952) study are commercially cultivated and harvested species, the question is strictly ecosystem oriented: what do the animals do to their environment? The answer requires the synthesis of information from the fields of biology, ecology, and sedimentary geology.

During the latter portion of the 1980s, Dame and Dankers (1988) and Dame et al. (1991a) investigated the exchange of materials between the intertidal mussel beds and the water column. These investigators used a 10-m long Plexiglas tunnel, BEST, to directly estimate the exchange rates of mussel beds holistically. Chlorophyll *a* was used as an estimator of phytoplankton. Using fishery statistics and other data, Dame et al. (1991) scaled-up their mussel bed exchange rates to the entire Western Wadden Sea system. They estimated that the mussel beds in the Western Wadden Sea could clear or filter the phytoplankton in about 20 days. The residence time of the water mass in the Wadden ranges from 5 to 15 days and is comparable to the system clearance rate for chlorophyll. From these calculations, it appears that mussels may control phytoplankton within this system, but this control is probably dependent on seasonal conditions and location.

ECOSYSTEM COMPARISONS

It is interesting that in the time since Officer et al. (1982) developed their relationship of bivalves and phytoplankton none of the systems they proposed as potential sites of filter feeder control other than San Francisco Bay have been investigated. Numerous other sites have been studied and an expanded comparison of the initial list developed by Smaal and Prins (1993) is presented in Table 12 and Figure 28. System clearance time is calculated as a function of bivalve filtration of the total water volume in an ecosystem and compared to water mass residence time. Also, bivalve dry body biomass is related to system water volume as suggested by Dame et al. (1991) in order to provide a system perspective of the relationships between bivalves and their food resource.

The relationship between water mass residence time and system clearance time by bivalves is only one aspect of the regulation of the water column by filter feeders. The amount and rate of the production of organic carbon by phytoplankton and how effectively that carbon is removed by bivalves is another aspect (Smaal and Prins 1993). These investigators developed a unitless measure they called filtration pressure. Filtration pressure (F/P) is the ratio of average system phytoplankton primary production (P) to average system filtration (F) or uptake of phytoplankton by bivalves (Table 13). Filtration pressure represents the proportion of phytoplankton primary production that is harvested by bivalves in a given system. Filtration pressure ranges from about 2% in Chesapeake Bay to greater than 100% in Sylt and Marennes-Oléron. In Sylt, the extra food is thought to be imported from the adjacent water body, while in Marennes-Oléron, resuspended microphytobenthos are implicated as an additional food resource. The Chesapeake Bay data are only for the current oyster population. Other types of filter feeders are not accounted for in this estimation of filtration pressure.

The preceding discussion focused mainly on systems where bivalve filter feeders are or were the major grazers in the ecosystem. In many of these systems, zooplankton were often considered an important grazing component, depending on the conditions. In a recent report, Mallin and Paerl (1994) reviewed the grazing of primary production by zooplankton in 13 estuarine and marine systems. They noted that on average zooplankton grazed between 4 and 70% of the primary production in the systems reviewed, but in some systems zooplankton grazing could potentially consume 100% of primary production during certain portions of the year. Only two of the systems surveyed by Mallin and Paerl (1994) were also in Table 12, Chesapeake Bay (White and Roman 1992) and Narragansett Bay (Verity 1987). Both of these systems currently have large open water areas with modest depth, conditions that support planktonic food webs. At the ecosystem level, it is not clear whether benthic bivalve filter feeders compete with zooplankton grazers for the phytoplankton resource or, as some mesocosm studies have shown, the bivalves actually filter out the zooplankton, particularly microzooplankton (Horsted et al. 1988, Riemann et al. 1988, 1990; Prins et al. 1995) and eliminate their competition.

Smaal and Prins (1993) suggest that the impact of bivalve filter feeders extends to various scales. These scales are the level of the bivalve bed, the estuary or bay, and the land–ocean interface. In those systems (Table 12) with low bivalve biomass-

TABLE 12
A Comparison of System Clearance Time by Bivalve Filter Feeders to Water Volume Residence Time

System	Symbol	Major Bivalve	Volume (10^6 m^3)	Biomass: Volume (g m^{-3})	Clearance Time (d)	Residence Time (d)	Clearance: Residence Ratio	Refs.
Askö	AS	Mussels	4,000	0.4	99	10^4	0.01	Kautsky and Wallentinus 1980
								Kautsky and Evans 1987
South San Francisco Bay	SSF	Various	2,500	2.5	0.7	11.1	0.06	Cloern 1982
Oosterschelde	ES	Mussels	2,740	3.1	3.7	40	0.09	Smaal et al. 1986
		Cockles						Dame et al. 1991
Chesapeake Bay, Past	CBO	Oysters	27,300	?	3.3	22	0.15	Newell 1988
Bay of Brest	BB	Various	1,480	7.4	2.8	16.7	0.17	Hily 1991
Marennes-Oléron Bay	MO	Oysters	675	4.2	2.7	7.1	0.38	Héral et al. 1988
								Bacher 1989
Ria de Arosa	RA	Mussels	4,335	1.6	12.4	23	0.54	Tenore et al. 1982
Western Wadden Sea	WW	Mussels	4,020	3.7	5.8	10	0.58	Dame et al. 1991
North Inlet	NI	Oysters	22	15.4	0.7	1.0	0.70	Dame et al. 1980
Narragansett Bay	NB	Clams	2,385		25	27	0.93	Kremer and Nixon 1978
Sylt, Eastern Wadden Sea	SY	Mussels	7.2	21.7	2.1	0.5	4.20	Asmus et al. 1990
Delaware Bay	DB	Oysters	19,420		1278	97	13.17	Biggs and Howell 1971
Chesapeake Bay, Present	CBP	Various	27,300		325	22	14.78	Newell 1988

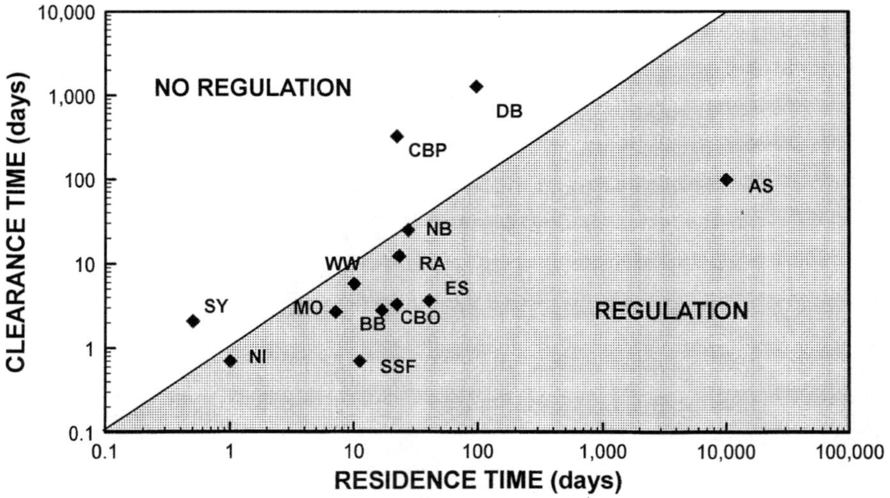

FIGURE 28. Graphical comparison of water mass residence time and clearance time in bivalve influenced ecosystems (expanded from Smaal and Prins 1993). Symbols defined in Table 12. Bivalves potentially regulate systems in shaded area where clearance time is shorter than residence time.

TABLE 13
Estimated Bivalve Filtration Pressure on Phytoplankton in Various Systems

System	Phytoplankton Primary Production *P* (gC m^{-2} year^{-1})	Bivalve Filtration *F* (gC m^{-2} year^{-1})	Filtration Pressure *F/P*	Refs.
Chesapeake Bay, Present	272	4.9	0.017	Ulanowicz and Tuttle 1992
North Inlet	265	40.9	0.150	Dame et al. 1991
Western Wadden Sea	248	45.5	0.183	Dame et al. 1991
Ria de Arosa	99	56.0	0.566	Tenore et al. 1982 Perez-Camacho 1991
Oosterschelde	190	112.0	0.589	Smaal and Van Stralen 1990 Prins and Smaal 1994
Sylt, Eastern Wadden Sea	73	137.0	1.877	Asmus et al. 1990
Marennes-Oléron	80	263.7	3.296	Bacher (per. comm.) Zurburg et al. 1994

Adapted from Smaal and Prins 1993.

to-water volume ratios, i.e., Killary, current Chesapeake Bay, Delaware Bay, and Narragansett Bay, the residence time of the water mass is less than that of the potential clearance time, and thus bivalve influences are probably limited to the level of the bed or community.

Systems with higher bivalve biomass to water volume ratios cleared larger volumes of water in less time than the water mass turned over (Smaal and Prins 1993). In these systems, i.e., Askö, Oosterschelde, Western Wadden Sea, Marennes-Oléron, etc., bivalve control of phytoplankton biomass was much more likely at the level of the bay or estuary when dry body bivalve biomass was in the range of 2 to 8 g m^{-3}, water mass residence times were long (sometimes called slow systems), and filtration pressure was high.

Phytoplankton biomass in short residence time systems (fast systems) could still be controlled by bivalves if their biomass:volume ratio is high, >8 g m^{-3}, as in North Inlet and Sylt. Thus, some fast systems function more like feedlots, with food being imported from the adjacent ocean to support high heterotrophic activity within their shallow tidal creeks and flats. Herman, as reported in Heip et al. (1995), has found that there is a very good inverse relationship between volume-specific biomass of bivalve suspension feeders and water mass residence time. Herman further speculates that food-limitation for system-wide bivalve suspension feeders may be a general property.

It is clear from the observations presented in Tables 12 and 13 that any model or conceptualization of the role of bivalve filtration in coastal waters should account for system water mass residence time, phytoplankton primary production or replacement time, and particle clearance rates or clearance time by the great majority of filtering organisms. These aspects should be fundamental components in any generic model of bivalve filter feeder dominated ecosystems as they represent the physical environment (water residence time), the primary producers (replacement time), and the grazers (clearance time).

Herman has recently proposed a simple model that includes these relationships (Heip et al. 1995). In Herman's model the rate of change of phytoplankton biomass per m^3 of water is described by:

$$\frac{dP}{dt} = P(\mu - m) - P(Cl_{ff})(B_{ff}) - \frac{P}{RT} + \frac{P_e}{RT}$$

where:

P = phytoplankton biomass (g m^{-3})
μ = growth rate of phytoplankton (day^{-1})
m = death rate of phytoplankton, resulting from causes other than grazing by bivalve suspension feeders (day^{-1})
Cl_{ff} = biomass-specific clearance rate of bivalve suspension feeders ($m^3g^{-1}d^{-1}$)
B_{ff} = biomass of bivalve suspension feeders (g m^{-3})
RT = water mass residence time (days)
P_e = phytoplankton concentration in advected water (g m^{-3})

Accepting a quasi-steady state, this relationship yields

$$B = \frac{(\mu - m)}{Cl_{ff}} = \left(\frac{P_e - P}{P \times Cl_{ff}} \right) \times \frac{1}{RT}$$

Realizing that Cl_{ff}, μ, m, and P_e/P are more or less the same in all systems, the relationships presented above describe that of the bivalve filter feeder's volume-specific biomass to water volume residence time. While not unreasonable, these relationships require more study of different ecosystems to confirm the hypothesized trends (Heip et al. 1995). These authors further argue that physical processes constrain bivalve filter feeders on a local scale, and primary production limits them on a system scale. Thus, suspension feeding bivalves will occur patchily (as reefs and beds), on a limited fraction of the total system surface.

As discussed earlier, Chesapeake Bay of the past was heavily dominated by bivalve filter feeding, but through overharvesting, disease, and changes in the surrounding landscape this system has changed to a planktonic one (Newell 1989). The switching from benthic–pelagic coupling to a planktonic system is but one way bivalves have been involved in changing the nature of a coastal system. In Ria de Arosa (Tenore et al. 1982) and Marennes-Oléron (Heral et al. 1988), we find the reverse case, where the artificial culture of dense populations of bivalves has moved these systems towards dominance by benthic bivalve filter feeding. From an ecosystem perspective, it would be interesting to know if such a phase shift can occur naturally and if it does, what the system advantages and disadvantages are. I suggest that pelagic planktonic grazing and benthic bivalve filter feeding are simply redundant components in the processing of suspended particulate material in coastal waters. However, when conditions favor benthic filter feeders, i.e., shallow flowing water, sufficient substrate, and adequate phytoplankton food supply, benthic bivalve filter feeders probably will dominate because their food chains are shorter, they take advantage of tidal energy subsidies to receive their food, and their longer life spans with greater stored biomass stabilizes a given ecosystem over longer time periods with a greater variety of environmental cycles.

REFERENCES

Alpine, S.E. and Cloern, J.E. 1992. Trophic interactions and direct physical effects control phytoplankton biomass and production in an estuary, *Limnol. Oceanogr.*, 37, 946–955.

Asmus, H. 1987. Secondary production of an intertidal mussel bed community related to its storage and turnover compartments, *Mar. Ecol. Prog. Ser.*, 39, 251–266.

Asmus, H. and Asmus, R.M. 1993. Phytoplankton-mussel bed interactions in intertidal ecosystems, in Dame, R.F., Ed., *Bivalve Filter Feeders in Estuarine and Coastal Ecosystem Processes*, Springer-Verlag, Heidelberg, 57–84.

Asmus, H., Asmus, R.M., Prins, T.C., Dankers, N., Frances, G., Maass, B., and Reise, K. 1992. Benthic-pelagic flux rates on mussel beds: tunnel and tidal flume methodology compared, *Helgol. Wiss. Meeresunters.*, 46, 341–361.

Asmus, H., Asmus, R.M., and Reise, K. 1990. Exchange processes in an intertidal mussel bed: a Sylt-flume study in the Wadden Sea, *Ber. Biol. Anst. Helgol.*, 6, 1–79.

Asmus, R.M. and Asmus, H. 1991. Mussel beds: limiting or promoting phytoplankton?, *J. Exp. Mar. Biol. Ecol.*, 148, 215–232.

Bacher, C. 1989. Capacité trophique du bassin de Marennes- Oléron: couplage d'un modèle de transport particulaire et d'un modele de croissance de l'huitre *Crassostrea gigas*, *Aquat. Living Resour.*, 2, 199–214.

Biggs, R.B. and Howell, B.A. 1984. The estuary as a sediment trap: alternate approaches to estimating filter efficiency, in Kenney, V.S., Ed., *The Estuary As A Filter*, Academic Press, New York, 107–129.

Butman, C.A., Fréchette, M., Geyer, W.R., and Starczak, V.R. 1994. Flume experiments on food supply to the blue mussel *Mytilus edulis* L. as a function of boundary-layer flow, *Limnol. Oceanogr.*, 39, 1755–1768.

Carlson, D.J., Townsend, D.W., Hilyard, A.L., and Eaton, J.F. 1984. Effect of an intertidal mudflat on plankton of the overlying water column, *Can. J. Fish. Aquat. Sci.*, 41, 1523–1528.

Chrzanowski, T., Stevenson, H., and Kjerfve, B. 1979. Adenosine 5′ triphosphate flux through the North Inlet marsh system, *App. Environ. Microbiol.*, 37, 841–848.

Cloern, J.E. 1982. Does the benthos control phytoplankton biomass in South San Francisco Bay?, *Mar. Ecol. Prog. Ser.*, 9, 191–202.

Cloern, J.E., Alpine, A.E., Cole, B.E. Wong, R.L.J., Arthur, J.F., and Ball, M.D. 1983. River discharge controls phytoplankton dynamics in northern San Francisco Bay estuary, *Estuarine Coastal Shelf Sci.*, 16, 415–429.

Cole, B.E., Thompson, J.K., and Cloern, J.E. 1992. Measurement of filtration rates by infaunal bivalves in a recirculating flume, *Mar. Biol.*, 113, 219–225.

Damas, D. 1935. Le role des organisms dans la formation des vases marines, *Ann. Soc. Geol. Belg.*, 58, 143–152.

Dame, R.F., Chrzanowski, T., Bildstein, K., Kjerfve, B., McKellar, H., Nelson, D., Spurrier, J., Stancyk, S., Stevenson, H., Vernberg, J., and Zingmark, R. 1986. The outwelling hypothesis and North Inlet, South Carolina, *Mar. Ecol. Prog. Ser.*, 33, 217–229.

Dame, R.F. and Dankers, N. 1988. Uptake and release of materials by a Wadden Sea mussel bed, *J. Exp. Mar. Biol. Ecol.*, 118, 207–216.

Dame, R.F., Dankers, N., Prins, T. Jongsma, H., and Smaal, A. 1991a. The influence of mussel beds on nutrients in the Western Wadden Sea and Eastern Scheldt estuaries, *Estuaries*, 14, 130–138.

Dame, R.F., Spurrier, J.D., Williams, T.M., Kjerfve, B., Zingmark, R.G., Wolaver, T.G., Chrzanowski, T.H., McKellar, H.N., and Vernberg, F.J. 1991b. Annual material processing by a salt marsh estuarine basin in South Carolina, USA, *Mar. Ecol. Prog. Ser.*, 72, 153–166.

Dame, R.F., Zingmark, R.G., and Haskin, E. 1984. Oyster reefs as processors of estuarine materials, *J. Exp. Mar. Biol. Ecol.*, 83, 239–247.

Dame, R.F., Zingmark, R.G., Stevenson, H., and Nelson, D. 1980. Filter feeding coupling between the water column and benthic systems, in Kennedy, V.S., Ed., *Estuarine Perspectives*, Academic Press, 521–526.

DiToro, D.M., O'Connor, D.J., and Thomann, R.V. 1971. A dynamic model of phytoplankton population in the Sacramento-San Joaquin delta, *Adv. Chem.*, 106, 131–180.

Doering, P.H. and Oviatt, C.A. 1986. Application of filtration rate models to field populations of bivalves: an assessment using experimental mesocosms, *Mar. Ecol. Prog. Ser.*, 31, 265–275.

Fréchette, M., Butman, C.A., and Geyer, W.R. 1989. The importance of boundary-layer flows in supplying phytoplankton to the benthic suspension feeder *Mytilus edulis*, *Limnol. Oceanogr.*, 34, 19–36.

Fréchette, M., Lefaivre, D., and Butman, C.A. 1993. Bivalve feeding and the benthic boundary layer, in Dame, R.F., Ed., *Bivalve Filter Feeders in Estuarine and Coastal Ecosystem Processes*, Springer-Verlag, Heidelberg, 325–369.

Gerritsen, J., Holland, A.F., and Irvine, D.E. 1994. Suspension-feeding bivalves and the fate of primary production: an estuarine model applied to Chesapeake Bay, *Estuaries,* 17, 403–416.

Grizzle, R.E. and Lutz, R.A. 1989. A statistical model relating horizontal seston fluxes and bottom sediment characteristics to growth of *Mercenaria mercenaria*, *Mar. Biol.*, 102, 95–105.

Grizzle, R.E. and Morin, P.J. 1989. Effect of tidal currents seston and bottom sediments on growth of *Mercenaria mercenaria*: results of a field experiment, *Mar. Biol.*, 102, 85–93.

Heip, C.H.R., Goosen, N.K., Herman, P.M.J., Kromkamp, J., Middelburg, J., and Soetaert, K. 1995. Production and consumption of biological particles in temperate tidal estuaries, *Oceanogr. Mar. Biol. Annu. Rev.*, 33, 1–149.

Héral, M. 1993. Why carrying capacity models are useful tools for management of bivalve culture, in Dame, R.F., Ed., *Bivalve Filter Feeders in Estuarine and Coastal Ecosystem Processes*, Springer-Verlag, Heidelberg, 455–477.

Héral, M., Deslous-Paoli, J.-M., and Prou, J. 1988. Approche de la capacité trophique d'un écosystème conchylicole, *J. Cons. Int. Explor. Mer*, Cm 1988/K, 22 pp.

Herman, P.M.J. 1993. A set of models to investigate the role of benthic suspension feeders in estuarine ecosystems, in Dame, R.F., Ed., *Bivalve Filter Feeders in Estuarine and Coastal Ecosystem Processes*, Springer-Verlag, Heidelberg, 421–454.

Herman, P.M.J. and Scholten, H. 1990. Can suspension-feeders stabilise estuarine ecosystems?, in Barnes, M. and Gibson, R.N., Eds., *Trophic Relationships in the Marine Environment*, Aberdeen University Press, Aberdeen, 104–116.

Hickman, R.W., Waite, R.P., Illingworth, J., Meredyth-Young, J.L., and Payne, G. 1991. The relationship between farmed mussels, *Perna canaliculus*, and available food in the Pelorus-Kenepuru Sound, New Zealand, *Aquaculture*, 99, 49–68.

Hilbert, D.W., Swift, D.M., Detling, J.K., and Dyer, M.I. 1981. Relative growth rates and the grazing optimization hypothesis, *Oecologia*, 51, 14–18.

Hily, C. 1991. Is the activity of benthic suspension feeders a factor controlling water quality in the Bay of Brest?, *Mar. Ecol. Prog. Ser.*, 69, 179–188.

Horsted, S.J., Nielsen, T.G., Tiemann, B., Pock-Steen, J., and Bjørnsen, P.K. 1988. Regulation of zooplankton by suspension feeding bivalves and fish in estuarine enclosures, *Mar. Ecol. Prog. Ser.*, 48, 217–224.

Jonge, V.N. de and Postma, H. 1974. Phosphorus compounds in the Dutch Wadden Sea, *Neth. J. Sea Res.*, 8, 139–153.

Jørgensen, C.B. 1943. On the water transport through the gills of bivalves, *Acta Physiol. Scand.*, 5, 297–304.

Kautsky, N. 1981. On the role of the blue mussel *Mytilus edulis L*, in The Baltic Ecosystem, Thesis summary, University of Stockholm, Sweden, 22 pp.

Kautsky, N. and Evans, S. 1987. Role of biodeposition by *Mytilus edulis* in the circulation of matter and nutrients in a Baltic coastal ecosystem, *Mar. Ecol. Prog. Ser.*, 38, 201–212.

Kautsky, N. and Wallentinus, I. 1980. Nutrient release from a Baltic *Mytilus*-red algae community and its role in benthic and pelagic productivity, *Ophelia*, Suppl. 1, 17–30.

Kemp, W.P. and Boynton, W.R. 1984. Spatial and temporal coupling of nutrient inputs to estuarine primary production: the role of particulate transport and decomposition, *Bull. Mar. Sci.*, 35, 242–247.

Kimmerer, W.J., Gartside, E., and Orsi, J.J. 1994. Predation by an introduced clam as the likely cause of substantial declines in zooplankton of San Francisco Bay, *Mar. Ecol. Prog. Ser.*, 113, 81–93.

Koseff, J.R., Holen, J.K., Monismith, S.G., and Cloern, J.E. 1993. Coupled effect of vertical mixing and benthic grazing on phytoplankton populations in shallow, turbid estuaries, *J. Mar. Res.*, 1–26.

Kremer, J. and Nixon, S. 1978. *A Coastal Marine Ecosystem*, Springer-Verlag, New York, 217 pp.

Loo, L. and Rosenberg, R. 1989. Bivalve suspension-feeding dynamics and benthic–pelagic coupling in an eutrophicated marine bay, *J. Exp. Mar. Biol. Ecol.*, 130, 253–276.

Lotsy, J.P. 1893. The food of the oyster, clam, and ribbed mussel, *Rep. U.S. Comm. Fish and Fisheries*, 19, 375–386.

Lund, E.J. 1957. A quantitative study of clearance of a turbid medium and feeding by the oyster, *Publ. Inst. Mar. Sci. Texas*, 4, 296–312.

Mallin, M.A. and Paerl, H.W. 1994. Planktonic trophic transfer in an estuary: seasonal, diel, and community structure effects, *Ecology*, 75, 2168–2184.

Maurer, D., Watling, L. and Keck, R. 1971. The Delaware oyster industry: a reality?, *Trans. Am. Fish. Soc.*, 100, 101–111.

McCrady, J. 1874. Observations on the food and reproductive organs of *Ostrea virginica*, with some account of *Bucephalus cuculus* nov. sp., *Proc. Boston Soc. Nat. Hist.*, 16, 170–192.

McNaughton, S.J. 1979. Grazing as an optimization process: grass-ungulate relationships in Serengeti, *Am. Nat.*, 113, 691–703.

Möbius, K. 1880. Die Auster und die Austerwirthschaft, *Rep. U.S. Comm. Fish.*, 683–751.

Møhlenberg, F. and Riisgård, H.U. 1979. Filtration rate, using a new indirect technique, in thirteen species of suspension-feeding bivalves, *Mar. Biol.*, 54, 143–147.

Moll, R.A. 1988. Phytoplankton in a temperate-zone salt marsh: net production and exchanges with coastal waters, *Mar. Biol.*, 42, 109–118.

Monismith, S.G., Koseff, J.R., Thompson, J.K. O'Riordan, C.A., and Nepf, H.M. 1990. A study of model bivalve siphonal currents, *Limnol. Oceanogr.*, 35, 680–696.

Morse, D.C. 1944. Some observations on the food and feeding of oysters in Chesapeake Bay, *Proc. Natl. Shellfish. Assoc.*, 55, 17–19.

Muschenheim, D.K. and Newell, C.R. 1992. Utilization of seston flux over a mussel bed, *Mar. Ecol. Prog. Ser.*, 85, 131–136.

Navarro, E., Ingelsias, J.I.P., Perez Camacho, A., Labarta, U., and Beiras, R. 1991. The physiological energetics of mussels (*Mytilus galloprovincialias*) from different cultivation rafts in Ria de Arosa (Galicia, N.W. Spain), *Aquaculture*, 94, 177–196.

Newell, R.I.E. 1988. Ecological changes in Chesapeake Bay: are they the result of overharvesting the American oyster, *Crassostrea virginica*?, in Lynch, M.P. and Krome, E.C., Eds., *Understanding the Estuary: Advances in Chesapeake Bay Research*, Chesapeake Research Consortium, Solomon's, Maryland, 536–546.

Nichols, F.H. 1985. Increased benthic grazing: an alternative explanation for low phytoplankton biomass in northern San Francisco Bay during the 1976–1977 drought, *Estuarine Coastal Shelf Sci.*, 21, 379–388.

Nichols, F.H. and Pamatmat, M. 1988. The ecology of the soft-bottom benthos of San Francisco Bay: a community profile, *USF&WS Biol. Rep.*, 85(7.19).

Nichols, F.H., Thompson, J.K., and Schemel, L.E. 1990. Remarkable invasion of San Francisco Bay by the Asian clam *Potamocorbula amurensis*. 2. Displacement of a former community, *Mar. Ecol. Prog. Ser.*, 66, 95–101.

Nienhuis, P.H. and Smaal, A.C. 1994. *The Oosterschelde estuary: A Case Study of a Changing Ecosystem* (reprinted from *Hydrobiogia* 282/283), Kluwer, Dordrecht, 597 pp.

Nixon, S.W., Oviatt, C.A., Rogers, C., and Taylor, K. 1971. Mass metabolism of a mussel bed, *Oecologia*, 8, 1–30.

Nowell, A.R.M. and Jumars, P.A. 1987. Flumes: theoretical and experimental considerations for simulation of benthic environments, *Oceanogr. Mar. Biol. Annu. Rev.*, 25, 91–112.

Odum, H.T., Burkhholder, P.R., and Rivero, J. 1959. Measurements of productivity of turtle grassflats, reefs and the Bahia Fosfrescente of southern Puerto Rico, *Publ. Inst. Mar. Sci.*, 6, 159–171.

Odum, H.T. and Odum, E.P. 1955. Trophic structure and productivity of a windward coral reef community on Eniwetok atoll, *Ecol. Monogr.*, 25, 381–411.

Officer, C.B., Smayda, T.J., and Mann, R. 1982. Benthic filter feeding: a natural eutrophication control, *Mar. Ecol. Prog. Ser.*, 9, 203–210.

O'Riordan, C.A., Monismith, S.G., and Koseff, J.R. 1993. An experimental study of concentration boundary layer formation over bivalve assemblages, *Limnol. Oceanogr.*, 38, 1712–1729.

Perez Camacho, A., Gonzalez, R., and Fuentes, J. 1991. Mussel culture in Galicia (N.W. Spain), *Aquaculture*, 92, 263–278.

Prins, T.C. 1996. Bivalve Grazing, Nutrient Cycling and Phytoplankton Dynamics in an Estuarine Ecosystem, Ph.D. dissertation, University of Wageningen, Wageningen, the Netherlands, 173 pp.

Prins, T.C., Escaravage, V., Smaal, A.C., and Peeters, J.C.H. 1995. Nutrient cycling and phytoplankton dynamics in relation to mussel grazing in a mesocosm experiment, *Ophelia*, 41, 289–315.

Prins, T.C. and Smaal, A.C. 1994. The role of the blue mussel *Mytilus edulis* in the cycling of nutrients in the Oosterschelde estuary, in Nienhuis, P.H. and Smaal, A.C., Eds., *The Oosterschelde estuary: A Case Study of a Changing Ecosystem*, Kluwer, Dordrecht, 413–429.

Reise, K. 1985. *Tidal Flat Ecology — An Experimental Approach to Species Interactions*, Springer-Verlag, Heidelberg, 191 pp.

Riemann, B., Nielsen, T.G., Horsted, S.J., Bjørnsen, P.K., and Pock-Steen. 1988. Regulation of phytoplankton biomass in estuarine enclosures, *Mar. Ecol. Prog. Ser.*, 48, 205–215.

Riemann, B., Sørensen, H.M., Bjørsen, P.K., Horsted, S.J., Jensen, L.M., Nielsen, T.C., and Søndergaard, M. 1990. Carbon budgets of the microbial food web in estuarine enclosures, *Mar. Ecol. Prog. Ser.*, 65, 159–170.

Rodhouse, P.G. and Roden, C.M. 1987. Carbon budget for a coastal inlet in relation to intensive cultivation of suspension-feeding bivalve molluscs, *Mar. Ecol. Prog. Ser.*, 36, 225–236.

Rogers, C.S. 1979. The productivity of San Cristobal reef in Puerto Rico, *Limnol. Oceanogr.*, 24, 342–349.

Rothschild, B.J., Ault, J.S., Goulletquer, P., and Héral, M. 1994. Decline of the Chesapeake Bay oyster populations: a century of habitat destruction and overfishing, *Mar. Ecol. Prog. Ser.*, 111, 29–39.

Sargent, M.C. and Austin, T.S. 1949. Organic productivity of an atoll, *Trans. Am. Geophys. Union*, 30, 245–249.

Smaal, A.C. and Prins, T.C. 1993. The uptake of organic matter and the release of inorganic nutrients by bivalve suspension feeder beds, in Dame, R.F., Ed., *Bivalve Filter Feeders in Estuarine and Coastal Ecosystem Processes*, Springer-Verlag, Heidelberg, 273–298.

Smaal, A.C., Verhagen, J.H.G., Coosen, J., and Haas, H.A. 1986. Interactions between seston quantity and quality and benthic suspension feeders in the Oosterschelde, the Netherlands, *Ophelia*, 26, 385–399.

Sommer, U. 1989. Nutrient status and nutrient competition of phytoplankton in a shallow hypertrophic lake, *Limnol. Oceanogr.*, 30, 355–346.

Sterner, R.W. 1986. Herbivores' direct and indirect effects on algal populations, *Science*, 231, 606–607.

Tenore, K.R., Boyer, L.F., Cal, R.M., Corral, J, Garcia-Fernandez, C., Gonzalez, N., Gonzalez-Gurrianan, E., Hanson, R.B., Iglesias, J., Krom, M., Lopez-Jamar, E., McClain, J., Pamatmat, M.M., Perez, A., Rhoads, D.C., de Santiago, G., Tiejen, J., Westrich, J., and Windom, H.L. 1982. Coastal upwelling in the Rias Bajas, NW Spain: contrasting the benthic regimes of the Rias de Arosa and de Muros, *J. Mar. Res.*, 40, 701–772.

Tilman, D. 1976. Ecological competition between algae: experimental confirmation of resource based competition theory, *Science*, 192, 463–465.

Thompson, J.K. and Nichols, F.H. 1988. Food availability controls seasonal cycle of growth in *Macoma balthica* (L.) in San Francisco Bay, California, *J. Exp. Mar. Biol. Ecol.*, 116, 43–61.

Ulanowicz, R.E. and Tuttle, J.H. 1992. The trophic consequences of oyster stock rehabilitation in Chesapeake Bay, *Estuaries*, 15, 257–265.

Van der Eijk, M. 1979. The Dutch Wadden Sea, in *Marine Production Mechanisms*, Cambridge University Press, Cambridge, U.K., 197–228.

Van Stralen, M.R. and Dijkema, R.D. 1994. Mussel culture in a changing environment: the effects of a coastal engineering project on mussel culture (*Mytilus edulis* [L.]) in the Oosterschelde estuary (SW Netherlands), in Nienhuis, P.H. and Smaal, A.C., Eds., *The Oosterschelde Estuary (The Netherlands: A Case Study of a Changing Ecosystem)*, Kluwer, Dordrecht, 359–379.

Verity, P.G. 1987. Abundance, community composition, size distribution and production rates of tintinnids in Narragansett Bay, Rhode Island, *Estuarine Coastal Shelf Sci.*, 24, 671–690.

Verwey, J. 1952. On the ecology of distribution of cockel and mussel in the Dutch Wadden Sea, their role in sedimentation and the source of their food supply, *Arch. Neerl. Zool.*, 10, 172–239.

Warwick, R.M. and Uncles, R.J. 1980. Distribution of benthic macrofauna associations in the Bristol Channel in relation to tidal stress, *Mar. Ecol. Prog., Ser.*, 3, 97–103.

White, J.R. and Roman, M.R. 1992. Seasonal study of grazing by metazoan zooplankton in the mesohaline Chesapeake Bay, *Mar. Ecol. Pro. Ser.*, 86, 251–261.

Wildish, D.J. 1977. Factors controlling marine and estuarine sublittoral macrofauna, *Helgol. Wiss. Meeresunters.*, 30, 445–454.

Wildish, D.J. and Kristmanson, D.D. 1984. The importance to mussels of the benthic boundary layer, *Can. J. Fish. Aquat. Sci.*, 41, 1618–1625.

Wildish, D.J. and Kristmanson, D.D. 1993. Hydrodynamic control of bivalve filter feeders: a conceptual view, in Dame, R.F., Ed., *Bivalve Filter Feeders in Estuarine and Coastal Ecosystem Processes*, Springer-Verlag, Heidelberg, 299–324.

Wildish, D.J. and Miyares, M.P. 1990. Filtration rate of blue mussels as a function of flow velocity: preliminary experiments, *J. Exp. Mar. Biol. Ecol.*, 142, 213–219.

Wildish, D.J. and Peer, D. 1983. Tidal current speed and production of benthic macrofauna in the lower Bay of Fundy, *Can. J. Fish. Aquat. Sci.*, 40(Suppl. 1), 309–321.

Winter, D.F., Banse, K., and Anderson, G.C. 1975. The dynamics of phytoplankton blooms in Puget sound, a fjord in the northwestern United States, *Mar. Biol.*, 29, 139–176.

Wright, R.T., Coffin, R.B., Ersing, C.P., and Pearson, C.P. 1982. Field and laboratory measurements of bivalve filtration of natural marine bacterioplankton, *Limnol. Oceanogr.*, 27, 91–98.

Zurburg, W., Smaal, A.C., Heral, M., and Dankers, N. 1994. *In situ* estimations of uptake and release of material by oysters in the Bay of Marennes-Oléron (France), in Dyer, K.R. and Orth, R.J., Eds., *Changes in Fluxes in Estuaries: Implications from Science to Management*, Olsen and Olsen, Fredensborg, 239–242.

6 System Metabolism and Nutrient Cycling

In this chapter, the role of bivalves in the ecosystem processes of metabolism and nutrient cycling is examined. In each case, the historical perspective for research in an area is presented along with the theoretical background. Methodologies are reviewed and examples utilizing bivalves are provided.

SYSTEM METABOLISM

One of the most important and time-proven measures of functional activity in ecological systems is the metabolic rate. System metabolism or metabolic rate is the sum of system respiration, primary production, and external exchanges. The rate of system metabolism is one of the few readily measurable emergent or holistic properties of ecosystems and allows systems to be categorized as autotrophic (net producer) or heterotrophic (net consumer) systems. The use of whole-stream (system) approaches to estimate system metabolism were first advocated by Odum (1956) and Odum and Hoskin (1958) for streams and estuaries, respectively. As with organisms, ecosystem metabolism can be estimated from temporal changes in the concentration of metabolic gases (O_2 and CO_2), and this approach has been utilized for many types of aquatic ecosystems (Kemp and Boynton 1980; Kenney et al. 1988; Marzolf et al. 1994).

The character of system metabolism is often dominated by subsystems within the ecosystem (Marzolf et al. 1994). In estuarine and coastal waters, beds of bivalve filter feeders (mussels, oysters, clams, etc.) may play a dominant role in the total metabolism of these systems. In the case of mussel and oyster beds, it is the high density of bivalve biomass that has been found to take up large quantities of O_2 (Nixon et al. 1971; Dankers et al. 1989; Dame et al. 1992) and allow these beds to dominate well-mixed tidal systems. In other reduced-flow systems, stratification of the water column allows benthic bivalves to regulate the heterotrophic nature of bottom waters (Cloern 1982, Alpine and Cloern 1992).

METHODS FOR MEASURING SYSTEM METABOLISM

As with individual organisms (Chapter 3), the major components of system metabolism, photosynthesis, and respiration can be estimated by the uptake and release of isotopic tracers for systems isolated in chambers (Chapter 7) or temporal changes in the concentrations of metabolic gases *in situ* or in micro- or mesocosms. The rates of photosynthesis and respiration are usually separated by measuring gas concentrations *in situ* over a 24 hour period (diurnal curves) or by using light and dark chambers (Odum 1956). The light component represents the sum of photosynthesis and respiration, while the dark component characterizes respiration. Subtracting the dark component from the light observations gives an estimate of photosynthesis. Both quiet water and flowing water upstream–downstream (see Chapter 5) time series of dissolved oxygen concentrations have found general acceptance in determining system metabolism.

In flowing waters, typical of bivalve-dominated environments, chambers reduce the realism of system metabolic measurements because mixing is reduced and the portion of the system observed is constrained (Kemp and Boynton 1980). The use of time series of oxygen concentrations to estimate system metabolism is based on the understanding that (1) during the day both benthic plants and phytoplankton release oxygen into the water column as a result of photosynthesis; (2) oxygen is taken from the water by organismic respiration and sometimes chemical oxygen demand; (3) there is an exchange of oxygen with the atmosphere in a direction depending on the saturation gradient; and (4) there may be an influx of oxygen in ground water and surface drainage (Odum 1956). The advantages of these methods are that the benthos remains undisturbed, all components of the system are included, and environmental conditions, i.e., water velocity, light, temperature, etc., are unaffected by the measurements. Disadvantages that are often encountered include: no true replication, confounded observations due to lateral flow, and corrections for air–water exchange of oxygen (Marzolf et al. 1994).

Chamber Measurements

A variety of investigators have utilized the chamber approach to obtain estimates of bivalve-dominated system metabolism. The role of clams in system metabolism was the focus of Murphy and Kremer (1985) and Doering et al. (1987), oyster community metabolism was investigated by Bahr (1976), and the influence of oyster biomass on benthic metabolism was examined by Boucher and Boucher-Rodoni (1988) and Boucher-Rodoni and Boucher (1990). In Colorado Lagoon, Long Beach, California, Murphy and Kremer (1985) used *in situ* opaque and clear hemispherical chambers (43 l) to measure aerobic respiration and chemical oxygen demand of the benthic environment dominated by the introduced clam *Mercenaria mercenaria*. These researchers extrapolated their chamber results to the entire lagoon (5.6 ha) and found that although microbial processes generally have been credited with being the most important component of benthic consumer communities, *Mercenaria* accounted for over 50% of the oxygen uptake within the system.

In a more realistic approach, Doering et al. (1987) used experimental mesocosms (13 m^3) to examine the effect of the hard clam *Mercenaria mercenaria* on the exchange of oxygen and carbon dioxide as well as inorganic nutrients across the sediment–water interface. The entire benthic component of the mesocosm was covered with a clear chamber (320 l) to follow material concentrations over time. The presence of clams at moderate densities increased the exchange of oxygen across the sediment–water interface and confirmed the observations of Murphy and Kremer (1985). Paradoxically, carbon dioxide production was not significantly different in chambers with and without clams. The investigators speculated that the lack of effect on carbon dioxide production may have been due to small sample size.

In studies by Doering et al. (1987), the respiratory quotient (CO_2:O_2) ranged from 0.26 to 3.29, but exhibited no treatment effect. The benthic respiratory quotient (RQ) is a combined index of aerobic and anaerobic respiration, and chemical oxidation of reduced products of anaerobic respiration. An RQ of 1.0 suggests that decomposition is mainly aerobic, or that reduced products of anaerobiosis are quickly oxidized chemically. A value greater than 1.0 implies that anaerobic production of CO_2 predominates over aerobic and chemical consumption of O_2. Thus, reduced products must either accumulate in the sediments or be exported from them (Hargrave and Phillips 1981). Although the RQ values in their study varied over an order of magnitude, indicating that aerobic and anaerobic respiration and chemical oxygen demand were not in balance in the short term, the long-term average RQ was about 1.0 and implies that these processes are roughly in balance on an annual basis (Doering et al. 1987).

Oxygen consumption by an intertidal oyster reef in Georgia was estimated by Bahr (1976), using samples incubated in plastic tubs. A layer of paraffin oil was used to seal the chambers from the atmosphere, and stratification was reduced through hand stirring. Oxygen consumption varied seasonally from 6 to 50 g m^{-2} day^{-1} with oysters, *Crassostrea virginica*, accounting for about 48% of the total uptake. No effort was made to partition the photosynthetic component.

In a series of studies, Boucher and Boucher-Rodoni (1988) and Boucher-Rodoni and Boucher (1990) investigated the respiratory metabolism of oyster beds in the Bay of Morlaix, North Brittany, France. Oxygen flux at the sediment–water interface was estimated *in situ* using hemispherical plastic chambers (40 to 50 l) with closed water circulation. Oysters, *Crassostrea gigas*, accounted for 26% of the annual oxygen uptake, and increasing oyster biomass increased oxygen uptake. These researchers also noted that oyster biomass influenced the coupling between oxygen consumption and inorganic nitrogen release, and the presence of oysters stabilized nitrogen release.

Free-Flow Measurements

In the published literature, only Nixon et al. (1971) have successfully measured the metabolism of a bivalve system using the upstream–downstream free-flow technique. Part of their success can be attributed to the location of the mussel bed they studied. It was located across the narrow, bottleneck opening than connects Bissel's Cove and the West Passage of Narragansett Bay, Rhode Island. The bed was completely exposed at low tide and was covered by flowing water at various depths on the ebbing and flooding tides. Oxygen concentrations were measured upstream and downstream of the bed using the Winkler method. Current velocities were estimated using drift bottles and dye along a transect between the upstream and downstream sample points. All estimates were made during the summer under daylight conditions, so photosynthesis was not estimated. Since diffusion to or from the atmosphere could influence oxygen concentrations in the water, a diffusion constant (reaeration constant) was derived from the equations of Edwards and Owen (1964) as checked by the floating dome method of Copeland and Duffer (1964). These investigators found that oxygen uptake was 0.2 g m^{-2} h^{-1} in quiet water and rose to 2.7 g m^{-2} h^{-1} in currents over 10 cm s^{-1}. Nixon et al. (1971) argued that the hyperbolic response of mussel bed oxygen consumption with response to increasing water velocity was indicative of the energy subsidy of the current serving as a limiting factor on community metabolism. Subsequent studies on experimental mussel beds with controlled water flow at velocities below 10 cm s^{-1} (Dankers et al. 1989) support the Nixon et al. (1971) interpretation for low flow conditions.

Tunnel Studies

The BEST tunnel approach (see Chapter 5) has been used to investigate the *in situ* metabolism of intertidal mussel beds in the Dutch Wadden Sea and oyster reefs in North Inlet, South Carolina (Dankers et al. 1989; Dame et al. 1992). Both studies utilized the diurnal approach to give estimates of day and night oxygen fluxes. The Dutch study also included an experimental flume design with constant water velocities and continuous upstream–downstream oxygen concentration monitoring with corrections for atmospheric exchanges. The mussel bed studies focused only on the summer season, and oxygen uptake rates of 1.2 to 3.9 g m^{-2} h^{-1} were observed (Dankers et al. 1989). The oyster reef investigations were uniformly distributed over the annual cycle and found a range of rates from a release of 1.5 g m^{-2} h^{-1} in winter to an uptake of 4.0 g m^{-2} h^{-1} in summer (Dame et al. 1992). The release of oxygen by the oyster reef in winter was attributed to the large growths of macroalgae on the beds and the reduced metabolism of the animals due to cold temperatures.

A summary of the system metabolism by a number of bivalve-dominated communities is provided in Table 14. Only the flux of oxygen per unit area is given because some reports did not provide bivalve biomass estimates. In general, systems with denser populations of bivalves have higher fluxes of oxygen and thus higher metabolism. As would be expected, rates are lower at lower temperatures in colder seasons, and production of oxygen can be observed when algal blooms occur on these beds.

NUTRIENT CYCLING

Bivalve molluscs not only remove materials from the water column or benthic–water interface, but they, as a result of feeding and metabolism, generate both particulate and dissolved materials that have an influence on benthic and pelagic habitats. Therefore, specific high quality organic constituents in the form of plankton, detritus, and amino acids are consumed as food, processed, deposited as feces or pseudofeces, or excreted as fundamental nutrients to the water column where they can be utilized by the plankton. Because bivalves often form dense assemblages or communities of organisms in shallow ecosystems, these animals may play an important role in the cycling of nutrients within these systems. In this section, the function of bivalves in nutrient cycling will be addressed. Specifically, the role these organisms play in the translocation, transformation, and remineralization of essential nutrients, as well as their part in feedback loops involving these elements will be examined.

TABLE 14
Observed Oxygen Fluxes in Bivalve Dominated Systems

System	Location	Oxygen Flux (mg m^{-2} h^{-1}) Min.	Max.	Ref.
Crassostrea gigas	France	1	160	Boucher and Boucher-Rodoni 1988
Crassostrea virginica	Georgia	6,000	50,000	Bahr 1976
	South Carolina	–1,500	4,000	Dame et al. 1992;
	Texas	500		Collier 1959
Mercenaria mercenaria	California	–40	800	Murphy and Kremer 1985;
	Rhode Island	1,830		Doering et al. 1987
Mytilus edulis	Rhode Island	200	2,700	Nixon et al. 1971
	The Netherlands	1,200	3,900	Dankers et al. 1989
Nearshore Benthos	Georgia	51	141	Hopkinson 1985

THEORETICAL BACKGROUND

Early studies by Liebig (1840) indicated that biologically essential elements (bio-elements or nutrients) could limit the production of organisms. At the level of the ecosystem, rather than a single limiting bioelement, it is often more complex interactions of elements that are limiting. The flux or cycling of essential materials is necessary for the continuity and stability of any living system and often provides a good indicator of metabolic activity and energy pathways through food webs. Consequently, the study of nutrient cycling is a major strategy in the analysis of ecosystems (Pomeroy 1970). There are a number of aspects to nutrient cycling including: (1) nutrient limitation of primary production; (2) recycling of nutrients and the mechanisms of recycling; (3) chemical complexity; and (4) stoichiometry (DeAngelis 1992). Nutrient limitation of primary production is common in all ecosystems and may be compensated for by internal recycling. In addition, the degree of system openness, as determined by the imports and exports of nutrients and when compared to the quantities of elements being recycled, is an important property at the ecosystem level. The cycle of each element is different from all others in its specifics, and some, like nitrogen, are very complex, involving a number of different compounds. Finally, chemical stoichiometry relates to the proportions of various elements, usually C, H, O, N, P, and S in a substance. These proportions are quite regular within groups of organisms and generally mean that one or more of these elements will be kept in short enough supply in the pool of available nutrients in an ecosystem to be limiting (DeAngelis 1992).

Initially, the role of macroconsumers in total ecosystem function was viewed from the context of energy flow. Kitchell et al. (1979) showed that large animals, e.g., bivalves, can influence nutrient cycling through physical and chemical processes not directly reflected by energy flow. Kitchell et al. (1979) summarize these processes into two general mechanisms of influence: translocation and transformation. In systems dominated by bivalve filter feeders, nutrients are moved from the water column to the benthos and back mainly through the pumping action of the animals. These translocated or retained nutrients can be distributed in ways that influence the total production rates of the system as a whole. Bivalves may also transform nutrients by changing their chemical complexity and by changing the particle size distribution in the water column. Normally, particulate organic matter is consumed, and dissolved inorganic materials are released as a part of bivalve metabolic processes. In addition, some bivalves may filter different portions of the suspended particulate size spectrum and this may also change material surface area-to-volume relationships. Finally, small particles in the water column are aggregated into the bodies of benthic bivalves and the relatively large size particles of their feces and pseudofeces.

A closer examination of nutrient cycling in ecosystems dominated by bivalves (Figure 29) indicates two basic types of feedback, negative and positive. Grazing or the consumption of primary producers by bivalves can be considered part of a predator–prey system where the bivalves benefit from eating the primary producers while the primary producers are reduced through consumption. As one component benefits and the other is negatively affected, this direct effect is referred to as negative feedback or deviation-counteracting. There is a general consensus in ecology that negative feedback regulation occurs to the degree that it normally keeps populations and communities from going completely out of control, although it may not always be strong enough to prevent sizable fluctuations (DeAngelis et al. 1986).

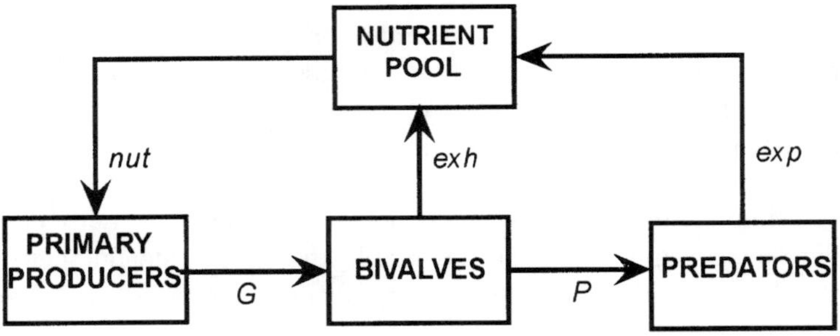

FIGURE 29. A simple diagram of nutrient cycling in bivalve dominated systems. *G* = grazing; *P* = predation; *exh* = excretion by bivalves; *exp* = excretion by predators; *nut* = nutrient uptake by primary producers.

The strong coupling of benthic filter feeding bivalves to phytoplankton in the water column via grazing (Chapter 5) is a negative feedback loop in the consumer–producer (predator–prey) mode. Such negative feedbacks should have a stabilizing influence on their ecosystems because they counteract deviations in the functional attributes of the system. Bivalve filter feeders may be especially stabilizing because (1) they are permanently in the ecosystem, (2) their filtration rates do not level off with increasing food availability, and (3) their biomass turns over at a low rate (Herman and Scholten 1990). Unlike predator–prey cycles in the plankton, the permanent presence of long-lived bivalves ensures that no time lags, a characteristic of negative feedback loops, occur between the appearance of food and the grazing activity of the animals. The bivalves are always waiting to feed, slowly starving during the winter and immediately grazing when primary production starts in the spring. The absence of a time lag may ensure that the phytoplankton cannot reach a level where effective grazing control becomes impossible (Herman and Scholten 1990).

Positive feedback is an indirect effect in bivalves systems where the utilize the nutrients in the primary producers and, in turn, the primary producers use the nutrients excreted by the bivalves. In essence, the bivalves are gardening their food resource, the phytoplankton. Both components benefit, but systems may respond to this feedback by reinforcing the change or effect in the direction of the deviation. Positive feedback amplifies changes and may result in boom or bust scenarios. DeAngelis et al. (1986) have made four generalizations about positive feedback: (1) the very existence of a positive feedback loop increases the complexity of the system through the network of nutrient flows and connections; (2) the positive feedback loop accelerates change within the system, i.e., bivalves short-circuit the food web, turning over nutrients more rapidly and making them available to the plankton for more production; (3) this type of feedback typically exhibits threshold effects where the mode of behavior of the system suddenly changes; and (4) systems that have major positive feedback loops may be fragile and the collapse or loss of one of the components can destroy the feedback loop, i.e., if the bivalve or the plankton components are lost from the system then the feedback loop and its features may be lost. Fragility may not always be a problem in bivalve systems as there are often redundant pathways and components, i.e., multiple bivalve and plankton species, that protect the system from collapse.

Nutrient Cycles

In the this section, nutrient cycling in bivalve-dominated systems will be discussed for specific elements. In the case of carbon and nitrogen, there are atmospheric (gaseous) forms that make these cycles more open, while in others, phosphorous and silicon, the cycles are more closed because they do not have an important gaseous form.

Carbon

Both inorganic and organic carbon are processed and stored by marine bivalves. Inorganic carbon is found in substantial amounts in the sea compared to the atmosphere and the land. Much of this carbon is present as dissolved inorganic carbon and forms a complex equilibrium reaction that buffers the pH of seawater (Valiela 1984). One component of this buffer system is calcium carbonate in the form of crystals in bivalve shells and structural components of other organisms, e.g., corals. In some groups, e.g., giant clams, their symbiotic relationship to zooxanthellae leads to the control of the carbonate buffer system within the clam and to an enhanced ability to lay down calcium carbonate crystals. As the cycling and storage of inorganic carbon take place on longer time scales than most ecosystem processes, this aspect of the carbon cycle will not be addressed here.

Photosynthetically active organisms can rapidly take up dissolved inorganic carbon and produce organic carbon that bivalves can utilize as food. This aerobic processing of organic carbon in marine systems with a significant bivalve component will be the focus of our discussions.

In many studies of the role of bivalves, energy flow and carbon cycling are used more or less synonymously (Baird and Milne 1981; Dame and Patten 1981). In the energy flow description of these systems, different forms of inputs and outputs, i.e., light energy, tidal energy, respiration, fecal production, etc., may not be distinguished. The oyster reef studied by Dame and Patten (1981) is a good example of this approach to carbon cycling (Figure 30). Oyster reefs have long been recognized as communities of organisms interacting with themselves and with their estuarine environments. These systems influence estuaries both physically, by removing suspended particulate material and changing current patterns, and biologically, by removing phytoplankton and producing large quantities of oyster biomass. In addition, the structure of the reef provides habitats for many sedentary and mobile organisms. The oyster reef as conceived by Dame and Patten (1981) has six major components: filter feeders, detritus, microbiota, meiofauna, deposit feeders, and predators. Their conceptualization is for an intertidal oyster reef in South Carolina with an average annual water temperature of 20°C. For simplicity and as supported by over 20 years of observations on the North Inlet, South Carolina system, their oyster reef is assumed to be at steady state (inputs = outputs). As in natural reefs, the epibenthic portions of the reef are dominated by the filter feeders, in this case *Crassostrea virginica* and *Brachidontes exustus*, and the rest of the system is sedimentary.

On the basis of Dame and Patten's (1981) calculations, this oyster reef consumes 4.2 gC m^{-2} d^{-1} or about 1514 gC m^{-2} $year^{-1}$ making it one of the most heterotrophic natural systems known. The filter feeders couple the reef to the water column, and this allows them to function both as grazers of phytoplankton and as suspended particulate detritus feeders. As there is little evidence that the filter feeders in the oyster reef utilize detritus as a food source, it is assumed that they simply deposit this material and it can then be utilized by other organisms in the reef.

In the Dame and Patten (1981) synthesis, the oyster reef is strongly heterotrophic, with food and waste carried in or taken away by the tides. Carbon deposited by the filter feeders is processed by a subsystem of organisms living in the sediments. An analysis of the carbon flows within the system (see Chapter 7) clearly shows the separation of the filter feeders from the rest of the system because there are no carbon feedbacks to the filter feeders from the rest of the system. Although there are numerous feedback pathways in the sediment portions of the system, only 11% of the material entering the reef is recycled with the remaining amount fluxing through the surface components of filter feeders, detritus, and predators. In the Dame and Patten (1981) analysis, the filter feeders clearly controlled the system, but the predators, mainly mud crabs, also exerted control over the detritus, microbiota, and meiofauna components.

The Dame and Patten (1981) conceptualization of carbon flow through an intertidal oyster reef was built from the bottom up using physiological and population data (see Chapters 3 and 4) on constituent organisms. Later studies by Dame et al.

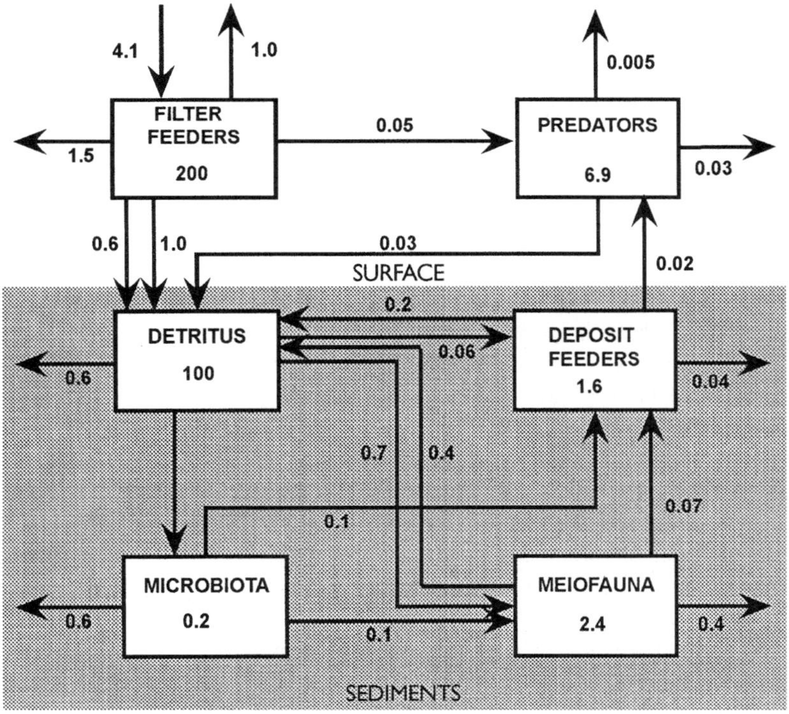

FIGURE 30. Carbon flow through an intertidal oyster reef in South Carolina. Flow = gC m^{-2} d^{-1}; states = gC m^{-2}. (Adapted from Dame and Patten 1981.)

(1989) used portable plastic tunnels to determine the input and output fluxes of the various forms of carbon, nitrogen, and phosphorus to an intertidal oyster reef in North Inlet. This group deployed the tunnel over a living oyster reef every 10.2 days (33 tidal cycles) over a single year and, using regression estimation techniques, computed an annual carbon uptake of 1200 gC m^{-2} $year^{-1}$. A value that is reasonably close to that estimated by Dame and Patten (1981) of 1514 gC m^{-2} $year^{-1}$. In the tunnel study, the oyster reef was considered to be a black box, and the fluxes of POC and DOC were observed. POC was taken up by the oyster reef on 95% of the observed flooding tides and released on 63% of the ebbing tides. The POC fluxes were probably the net result of both biofiltration and the physical processes of sedimentation and resuspension on the oyster reef. The DOC component was released on 63% of both flooding and ebbing tides. This constituent can be taken up by filter feeders as a nutritional subsidy in the form of amino acids (Manahan et al. 1982; Stephens 1982; Wright 1982) and can be excreted as urea, amino acids, and uric acid (Hammen et al. 1966). In addition, other organisms on the reef, such as bacteria and algae, can take up and release DOC (Valiela 1984), and increased water velocities may increase the release of DOC from benthic systems (Boynton et al. 1981).

In conclusion, the Dame and Patten (1981) carbon flow description of the oyster reef was confirmed by the black box–tunnel field measurements (Dame et al. 1989).

These systems are highly heterotrophic and filter large quantities of POC from the water column and release DOC back to the surrounding environment. As generalized by Kitchell et al. (1979), oyster reefs translocate and transform large quantities of carbon in estuaries and can be considered high speed estuarine processors.

At a slightly higher level of complexity, but still historically early in carbon budget descriptions, Warwick et al. (1979) developed a carbon flow model of a mud flat in the Lynher estuary, Cornwall, U.K. Their initial conceptualization was developed from steady-state carbon budgets for major groups including: diatoms, meiofauna, deposit feeders, suspension feeders, and worms driven by external variables of particulate carbon, bacteria, phytoplankton, fish, and birds (Figure 31). From a bivalve perspective, their model is unique because it includes both suspension feeding bivalves and deposit feeding bivalves. The suspension feeders were dominated by *Mya* and *Cardium* = *Cerastoderma*, while the deposit feeders were represented by *Macoma* and *Scrobicularia.* The steady-state model highlighted the lack of information about meiofauna and small annelids in particular. The steady-state version of the model was further developed into a dynamic simulation model that realistically mimicked the deposit feeders. As with the steady-state model, this version also reflected the lack of good data on the kinetics of feeding by meiofauna and macrofauna, especially in relation to food concentrations and assimilation rates. As with the preceding carbon flow representations, there were no positive feedback loops and only tacit negative feedbacks were present.

A steady-state carbon flow model was constructed for the Ythan Estuary in Scotland by Baird and Milne (1981). This model also included both suspension feeding and deposit feeding bivalves represented by *Mytilus edulis* and *Macoma balthica*. *Mytilus* appears to consume a major portion of the phytoplankton produced in the system, although a significant amount is exported to the sea. *Macoma* is thought to utilize benthic diatoms as well as particulate organic matter from macrophytes. The production of macrofauna appears to be at least twice that of predator demand and partially explains increases in bird populations. As with the Warwick et al. (1979) study, meiofauna need further investigation in this system. Although it is not clearly shown, this model is the first attempt to include inorganic nutrients and their recycling in a system dominated by bivalves.

As described earlier in Chapter 5, Killary Harbour is one of the principal areas in Ireland for the intensive suspended cultivation of mussels. Rodhouse and Roden (1987) developed a carbon flow model for this system emphasizing the roles of natural shore mussel populations, raft cultured mussels and herbivorous zooplankton (Figure 32). This synthesis focused on the potential roles of the various suspension feeding components and whether one or the other was enhanced. The study recognized that the water column within the estuary produced insufficient phytoplankton to support the size of the suspension feeder populations and that phytoplankton must be imported from outside the inlet. In their conceptualization, Rodhouse and Roden (1987) assumed that increasing mussel raft production would decrease the zooplankton component of the estuary.

Ulanowicz and his colleagues have developed a series of carbon flow models for Chesapeake Bay that include oysters and bivalves as major components (Baird

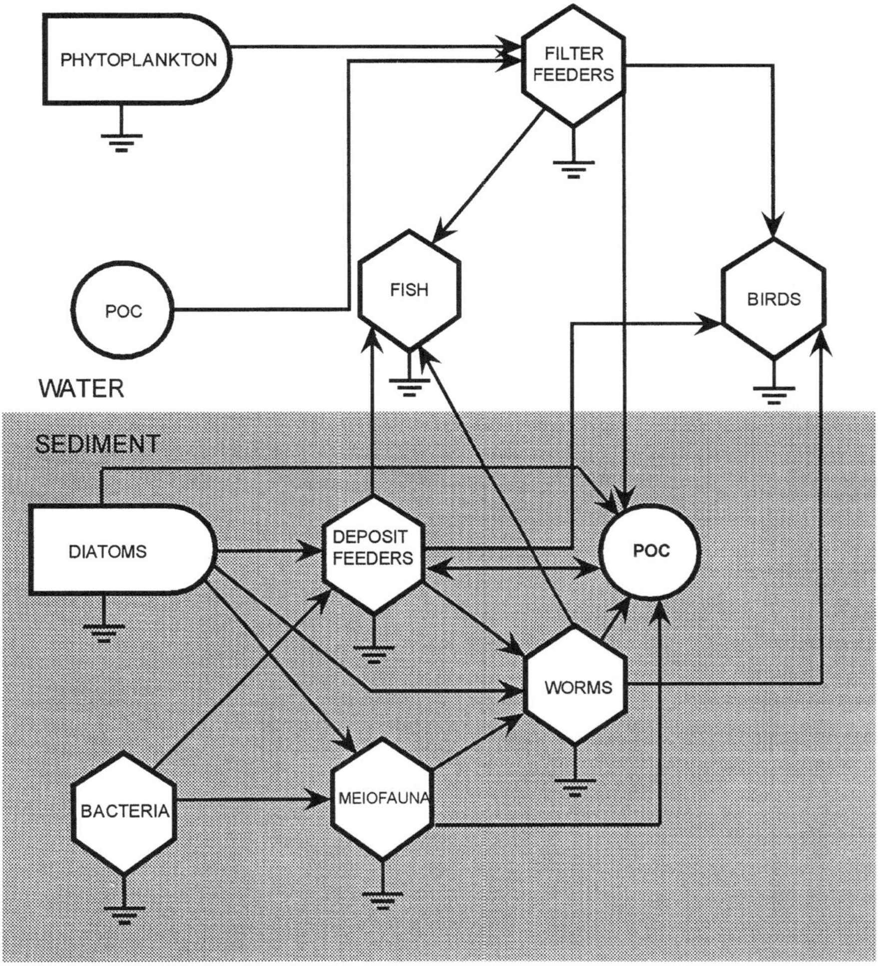

FIGURE 31. A conceptual model of carbon flow in Lynher mud-flat system in Scotland showing the dual role of suspension feeding and deposit feeding bivalves. (Adapted from Warwick et al. 1979.)

and Ulanowicz 1989; Baird et al. 1991; Ulanowicz and Tuttle 1992). Their original conceptualization of carbon flow in Chesapeake Bay (Baird and Ulanowicz 1989) was composed of 36 components that they thought were significant. These components included water column and benthic primary producers, decomposers, oysters and other filter feeders, deposit feeders, numerous different fishes, and predators. An aggregated form of this model was subsequently developed that had 13 components (Figure 33) and was used to address specific questions about the role of oysters within the Chesapeake Bay ecosystem (Ulanowicz and Tuttle 1992).

The Ulanowicz and Tuttle (1992) depiction of carbon flow through the Chesapeake Bay ecosystem is similar to that of Dame and Patten (1981) for an oyster reef in that the filter feeders (oysters) receive no feedback flows. This approach has also

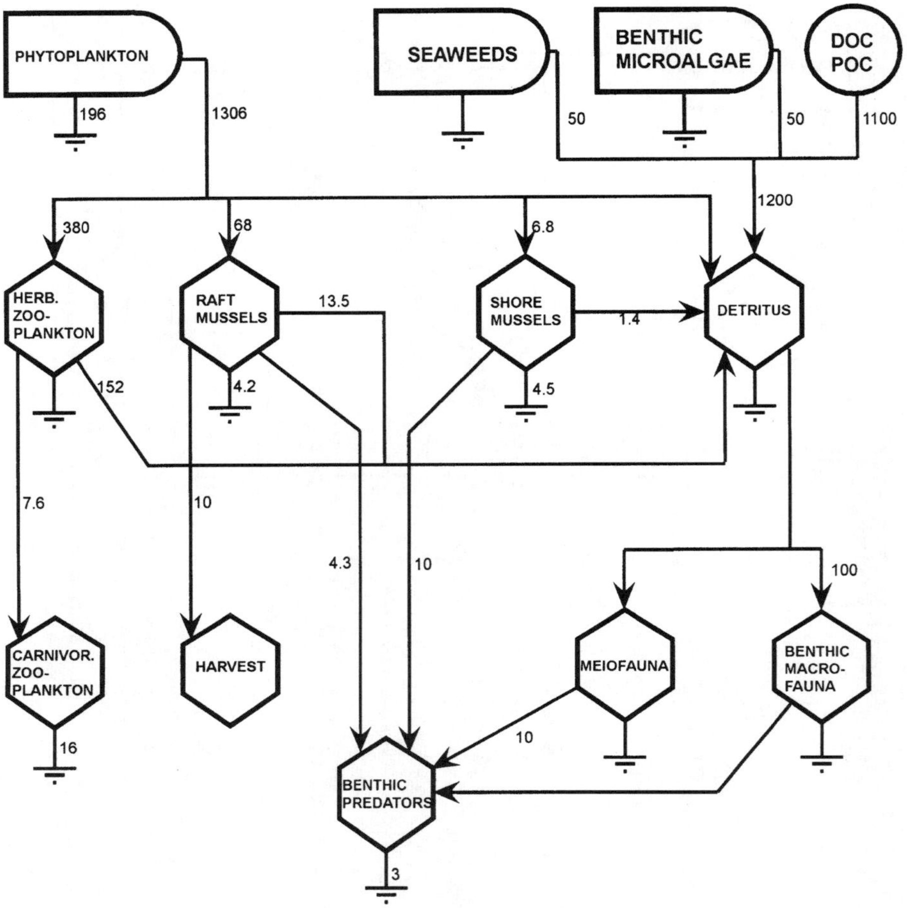

FIGURE 32. A carbon flux model of Killary Harbour, Ireland, showing the interrelationships of pelagic zooplankton, raft mussels, and shore mussels, the major grazers, in the system. Units in tons C per year. (Adapted from Rodhouse and Rodin 1987.)

been used for systems dominated by mussels feeding mainly on phytoplankton (Baird and Milne 1981; Rodhouse et al. 1981).

In controlled mesocosm experiments (Chapter 7) that were designed to simulate conditions in shallow unstratified coastal waters, e.g., Narragansett Bay, Doering et al. (1986) examined the effects of the clam, *Mercenaria mercenaria*, on carbon cycling. In their studies, these authors used radioactively labeled carbon, ^{14}C, to examine systems with and without clams. They found that chambers with clams had greater system production, sedimentation of particulate carbon, and benthic remineralization. The presence of *Mercenaria* did not change the standing stock of phytoplankton, but the possible positive feedback of remineralized nutrients may have offset the negative effects of grazing. These authors point out that the residence time of water in systems subjected to bivalve filtering could determine the importance of these feedbacks. They further suggested that the effects of filter feeders on system

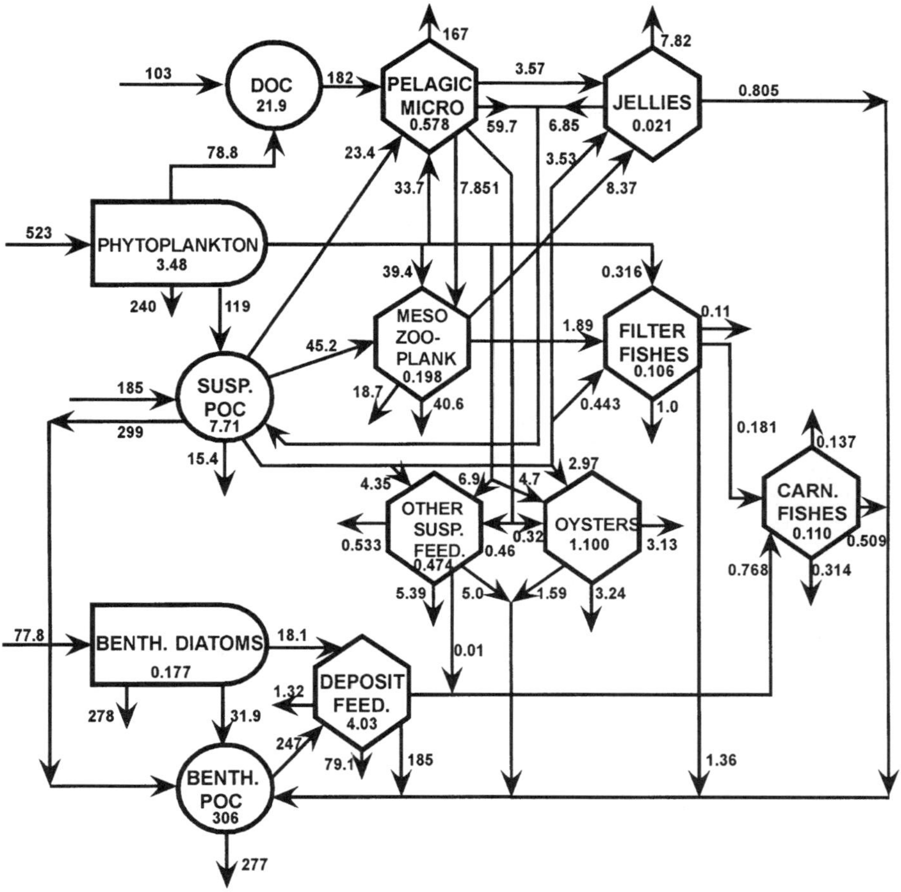

FIGURE 33. Carbon flow through Chesapeake Bay. Flows = gC m^{-2} year^{-1}; states = gC m^{-2}. (Adapted from Ulanowicz, R.E. and Tuttle, J.H. 1992.)

function vary with density and that the variations are not only quantitative, but qualitative.

However, Newell and Field (1983) described carbon flow for a kelp bed system with a significant component of filter feeding mussels, *Aulacomya ater*, that were potentially consuming 9% of their carbon input in the form of bacteria generated from the decomposition of kelp detritus (Figures 34a and 34b). This positive feedback of carbon via the decomposer loop and back to the bivalve filter feeders may be more common to mussel-dominated systems, as Wright et al. (1982) have shown that mussels can filter and consume smaller (bacteria size) particles than oysters. On the basis of the available information, carbon flow in oyster-dominated systems does not appear to have any clearly identifiable positive feedback flows.

To summarize, the role of bivalves in carbon cycling can be important when large amounts of organic carbon are being processed. In shallow low flow systems, it is probable that removing oxygen from the water and releasing carbon dioxide in

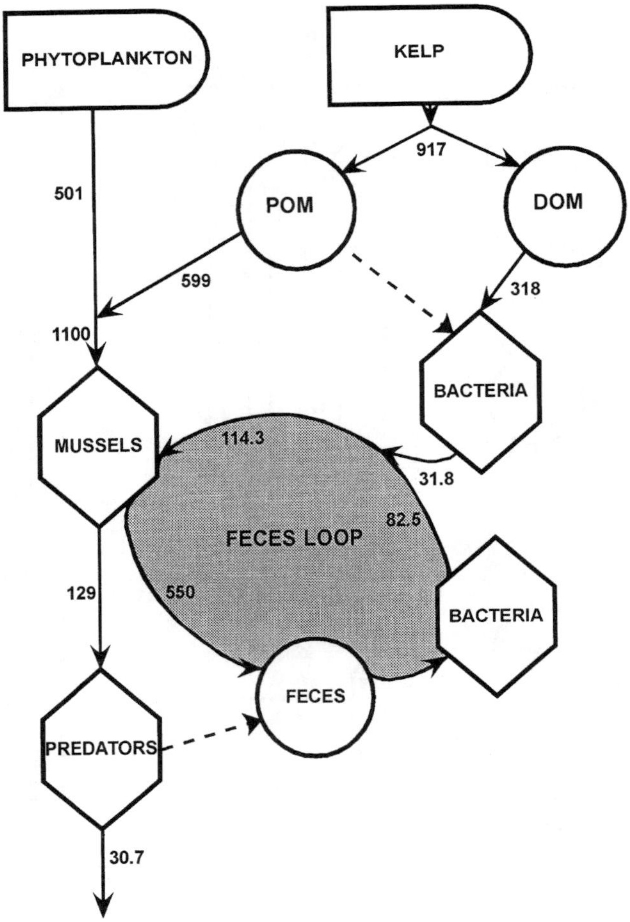

FIGURE 34a. Carbon flow and cycling through a kelp bed community off South Africa. Flows = gC m^{-2} $year^{-1}$. (Adapted from Newell, R.C. and Field, J.G. 1983.)

return are not only going to influence environmental conditions but will also set up potential positive feedbacks between the bivalves and other primary producers. The clearly defined negative feedback loops in conjunction with positive feedback loops will probably lead to more stable ecosystems when bivalves are present.

Nitrogen

Nitrogen is an element that undergoes numerous transformations in addition to state changes as it cycles through the environment. After carbon, nitrogen may be the most important element to living organisms, especially those that live in marine and coastal environments where nitrogen may be limiting (Sprent 1987). Compared to carbon that has only one inorganic form, nitrogen has several that play important roles in environments where bivalves may be dominant components. It is probably

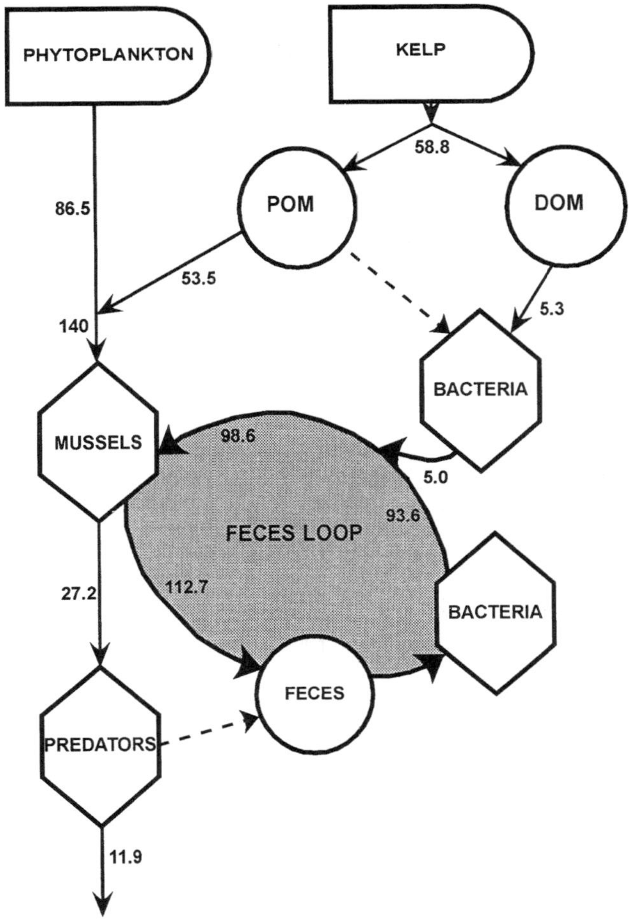

FIGURE 34b. Nitrogen flow and cycling through a kelp bed community off South Africa. Flow = gN m^{-2} $year^{-1}$. (Adapted from Newell, R.C. and Field, J.G. 1983.)

this diversity of forms and states that has retarded the development of nitrogen budgets for ecosystems in general and coastal systems in particular (Nixon and Pilson 1983). Thus, it is not surprising that observation of nitrogen processing by bivalves in general is incomplete for most systems studied.

In many systems, input of new nitrogen, i.e., nitrogen from fixation, is not sufficient to support the needs of primary production. In these cases, particulate organic nitrogen is recycled by the release of inorganic nitrogen by heterotrophic consumers, e.g., oysters. Usually this inorganic nitrogen is ammonium and often this form of nitrogen is preferentially taken up by phytoplankton. Bivalves may also release dissolved organic nitrogen in the form of amino acids and urea (Hammen et al. 1966) that can also be used by phytoplankton (McCarthy 1980). Thus, in the nitrogen cycle in coastal waters, benthic bivalves can short-circuit pelagic nutrient processing and rapidly recycle nitrogen. The nitrogen feedback loop between the bivalves and the phytoplankton is positive because both the phytoplankton and the bivalves benefit from the process.

Several studies have focused on the role of mussels from different environments in the processing of nitrogen. In the Baltic Sea adjacent to Askö, Kautsky and Wallentinus (1980) determined the fluxes of nitrogen from a *Mytilus*–red algae community utilizing *in situ* plastic bags. Their observations showed that nutrient release by the mussels greatly exceeded demand by the benthic algae and may explain why the algae reach maximum growth after nutrient concentrations in the surrounding water column are depleted by the spring phytoplankton bloom. These investigators further speculate that the removal of phytoplankton by the mussels also positively influences the growth of the red algae by increasing light availability to the benthos. For the Askö system, the mussels produce via recycling several orders of magnitude greater nutrients than can be accounted for from land runoff.

Because of the magnitude of particle removal by the mussels, Kautsky and Evans (1987) conducted *in situ* studies on the production of feces and pseudofeces by *Mytilus edulis* in the Askö area. They found that the concentrations of carbon and nitrogen in the *Mytilus* biodeposits were higher than that of sediments deposited through natural physical means in the same area. Thus, these investigators concluded that mussels remove and initiate the sedimentation of small suspended particles of high organic content that would otherwise stay suspended in the water column. They argue that *Mytilus* occupies an important position as a connecting link between the pelagic and benthic components of the coastal ecosystem by increasing sedimentation by as much as 10% and enhancing the recycling of nutrients to the primary producers in both the pelagic and benthic habitats. In essence, the mussels strengthen the coupling between the benthos and the water column.

In a very different environment, Jordan and Valiela (1982) investigated the nitrogen budget of the ribbed mussel, *Geukensia demissa*, and its significance in nitrogen flow in a New England salt marsh. Working initially at the population level, they developed a nitrogen budget for the marsh mussel (Figure 35). Of the nitrogen filtered from the water column by the mussel population, slightly less than half was deposited as feces and pseudofeces. Of the nitrogen absorbed, 55% was excreted as ammonia. At the level of the entire marsh–estuarine ecosystem (Figure 36), the mussels filtered 1.8 times the particulate nitrogen exported from the salt marsh by tidal flushing. On the basis of these comparisons, the marsh mussels released more ammonia into the system than any other group of organisms. Thus, marsh mussels increase retention of nitrogen within the marsh by filtering particulate nitrogen from the water column, depositing it as biodeposits, and storing it as long-lived mussel biomass. As nitrogen limits productivity in this marsh, increased retention and remineralization of nitrogen by marsh mussels may ultimately enhance the productivity of the salt marsh. Indeed, an experimental study by Bertness (1984) in a similar New England salt marsh showed that *Spartina alterniflora* grass with large clumps of marsh mussels had primary production higher than grass without mussels. Bertness (1984) found that the sediments surrounding the mussels had higher concentrations of inorganic nitrogen and attributed this fact to mussel excretion and feces production. Thus, the nitrogen retained in the system may go to another form of plant other than phytoplankton and not to water column microalgae that would

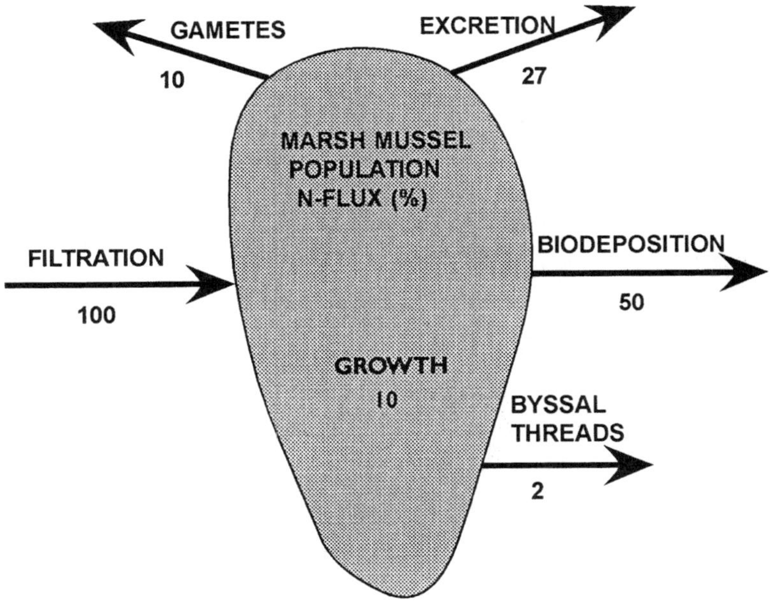

FIGURE 35. The relative percentage nitrogen flux in a population of the marsh mussel, *Geukensia demissa* in a New England salt marsh. (Data from Jordan and Valiela 1982.)

complete a direct positive feedback loop. However, it is possible that the positive feedback to the mussels from increased *Spartina* growth may result in a better structural habitat for the marsh mussels.

The first direct evidence that bivalve filter feeders could influence nitrogen fluxes at the ecosystem level came from Nixon et al. (1976). In their study of the Bissel Cove salt marsh embayment in Narragansett Bay, a bed of *Mytilus* was observed to excrete 4 to 5 *mM* ammonia m^{-2} h^{-1} at 15°C. These authors hypothesized that blue mussels in this system import large quantities of suspended particulate nitrogen and export large amounts of ammonia to the adjacent offshore water.

It was not until the application of portable plastic tunnels (Dame et al. 1984) and domes (Boucher and Boucher-Rondoni 1988) that *in situ* measurements of nitrogen uptakes and releases were conducted on bivalve systems. A major objective of each of these studies was to observe the material fluxes across bivalve beds in as realistic manner as possible.

The initial tunnel work was conducted on intertidal oyster reefs, *Crassostrea virginica*, in North Inlet, SC and is described in Chapter 5. The simultaneous fluxes of various forms of nitrogen were measured every 10.2 days for a year (Table 15). A striking observation of this study was the considerable variation in material fluxes from one observed tidal cycle to the next. This variability was attributed to tide-to-tide differences in ebb and flood velocity patterns. In addition to the general net flux pattern of uptake of particulates and release of dissolved materials, the observed fluxes were almost always an order of magnitude greater than scaled-up estimates from laboratory data. This latter finding supports the argument that more nutrient

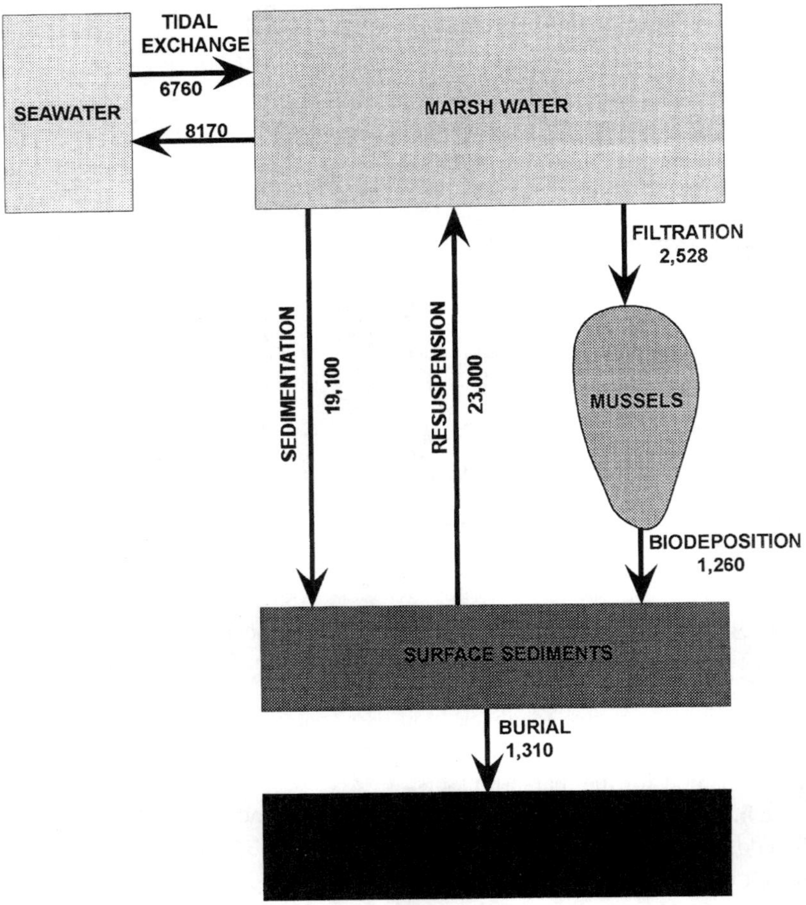

FIGURE 36. Nitrogen flux in Great Sippewissett salt marsh, depicting the importance of marsh mussels in the ecosystem. (Adapted from Jordan and Valiela 1982.)

processing is also taking place in the adjacent sediments and by other organisms in the reef community. The high flux rates observed in this study strongly suggest a major role for oysters in ecosystems where they are abundant. When scaled up to the marsh–estuarine basin, the oyster reefs are by comparison the largest and only significant source of ammonium within this part of the North Inlet system (Dame et al. 1991b). In North Inlet, the turnover time for ammonium as calculated from oyster reef release is about half that of the water mass residence time. This relationship implies that the oyster reefs are controlling ammonium concentrations in North Inlet.

In the large tidal range estuaries of North Brittany, France, Boucher and Boucher-Rodoni (1988) and Boucher-Rodoni and Boucher (1990) investigated the role of the oyster, *Crassostrea gigas*, in nutrient fluxes. Using relatively low densities of oysters in domes, they found that the oysters contributed about 15 to 40% of the ammonium and urea within the chambers. They also noted that the actual material fluxes due

TABLE 15
Yearly Nitrogen and Phosphorus Budgets for an Intertidal Oyster Reef in North Inlet, SC

Constituent	Flood Tide	Ebb Tide	Net
Total dissolved nitrogen	–116*	91	–25
Ammonium	–58*	–67*	–125*
Nitrite + nitrate	0.3	–1.3	–1.0
Dissolved organic nitrogen	–136*	263*	127
Particulate nitrogen	472*	–250	222
Total nitrogen	275*	–86	189
Orthophosphate	1.4	–9.1	–7.7
Particulate phosphate	156*	–86	70
Total phosphorus	136*	–38*	98*

Note: Units in gN m^{-2} $year^{-1}$ or gP m^{-2} $year^{-1}$

* Flux significant at 5% level; +: uptake (no sign indicated); –: release.

After Dame, R.F., Spurrier, J.D., and Wolaver, T.G. 1989.

to the oysters were always greater than potential estimates. In these dome studies, some nitrate production was observed and these researchers suggest that this indicated occasional nitrification, i.e., the transformation of ammonia to nitrate.

The nitrogen excreted by bivalves and released by organically enriched adjacent sediments can have a major influence on the microalgae in bivalve culture facilities (Robert et al. 1982; Maestrini et al. 1986). These researchers found that the microalgae seem to have adapted to the high concentrations of ammonium, nitrate and organic nitrogen in these culture ponds. Normally high concentrations of ammonium inhibit the uptake of other nitrogen sources by algae, but the threshold of ammonium inhibition is much higher in these pond algae than in pelagic or neritic species, allowing them to assimilate other sources of nitrogen simultaneously with ammonium.

A number of studies dealing with the intensive cultivation of mussels and oysters have shed more light on the role of bivalves in the processing of nitrogen in the sediments (Robert et al. 1982; Kaspar et al. 1985; Feuillet-Girard et al. 1988). In all of these studies, the sediments in bivalve culture areas or ponds were compared to areas without bivalve culture. In early studies on the pond culture of *Crassostrea gigas* in the Marennes-Oléron area of France, culture ponds with oysters had very high concentrations of dissolved nitrogen compared to ponds without oysters. In addition, there was higher microalgae production in the ponds with oysters. Subsequent studies in Marennes-Oléron Bay found that the sediments in oyster cultivation areas accumulated 6 times as much nitrogen as those without adjacent oysters. The fluxes of nitrogen from the sediments were higher in cultivation zones than noncultivation zones. These authors concluded that oyster cultivation induces nitrogen fixation in the sediments above noncultivated levels. In their concept of nitrogen cycling in the oyster culture areas of Marennes-Oléron Bay, Feuillet-Girard et al. (1988) concluded that oysters accounted for about 50% of the nitrogen reaching the sediments from the water column, but the influence of the oysters on dissolved nitrogen concentrations was minimal.

In addition to increasing nitrogen fluxes to the sediments, the enriched biosediments near bivalve culture plots also influence a number of biogeochemical processes in the sediments. Kaspar et al. (1985) compared the effects of the culture of the green-lipped mussel *Perna canaliculus* on the nitrogen cycle in Marlborough Sounds, New Zealand. The nitrogen cycle for the mussel farm is described in Figure 37. As in the French studies, they found that the organic nitrogen pool in the sediments beneath the mussel farm was larger than at a reference site. The nitrate and nitrite pools were similar in the sediments at both sites, and the ammonium pool was about twice as high in the culture sediments as in the reference sediments. As in other studies, the mussels and the mussel farm are envisioned as a large filter that forms fast-sedimenting pellets from planktonic particles that accumulate as organic nitrogen in the sediments. The larger ammonium pool of the mussel farm sediments was attributed to higher ion-exchange capacity of the sediments due to the increased organic matter content. Thus, the mussel-induced increased sedimentation directly stimulated the remineralization of organic nitrogen in the sediments. These authors also speculate that by producing organically enriched sediments, the mussels are indirectly stimulating denitrification in the deeper anaerobic sediments, i.e., the conversion of nitrate to nitrogen gas. Denitrification by anaerobic sediments adjacent to bivalve communities has been argued by some (Roger Newell, personal communication) as a way of removing nitrogen from extensively nutrient loaded or polluted estuaries. Kaspar et al. (1985) also point out that in addition to increasing nitrogen turnover in the system, mussels that are harvested are a nonreturned sink of nitrogen that may lead to system nitrogen depletion and limitation. To summarize, (1) bivalves enhance the movement of organic nitrogen to the sediments where it decomposes; (2) decomposition in the aerobic surficial sediments yields ammonium (mineralization) and nitrate and nitrite (nitrification); (3) in the deeper anaerobic sediments the nitrate is converted into nitrogen gas (denitrification).

In order to directly assess the role of oyster reefs on nutrient cycling in a group of tidal creeks, Dame and Libes (1993) developed a design to experimentally manipulate these systems. In their experiment, nutrient concentrations in creeks with oysters were compared to creeks with oysters removed using a BACI (before, after, control, incident) design (Underwood 1994). The amount of oyster biomass to water volume in each creek was standardized to that of the North Inlet system (3.8 g m^{-3}) and the bank-full volume of each creek was determined by hypsometric characterization. After a before-manipulation calibration run of 1 month, oysters were removed from half of the creeks and the daily concentrations of nutrients were monitored for 2 months. With regard to nitrogen, ammonium, and total nitrogen concentrations were found to be significantly higher in creeks with oysters when compared to creeks without oysters. These data clearly and directly support the

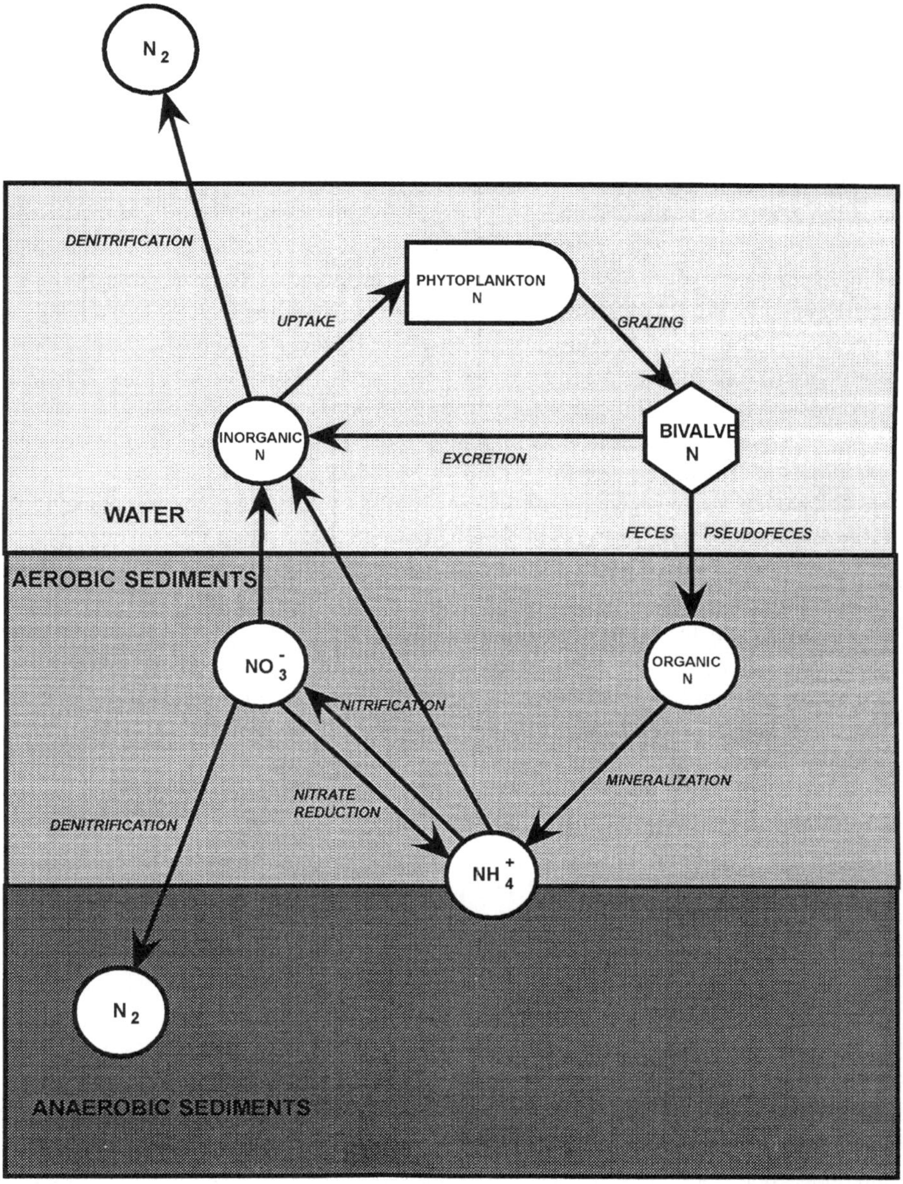

FIGURE 37. The nitrogen cycle in a bivalve filter feeder-dominated system.

previous suggestions that dense populations of bivalves can and do significantly influence the amounts and types of nitrogen in tidal waters. In addition, this preliminary study supports the idea that bivalves may play a major role in material cycling in these systems.

Phosphorus

Phosphorus is one of the essential elements in living organisms and plays a fundamental role in the metabolic processes of energy transfer, both respiration and photosynthesis. In marine environments, phosphorus is found in living organisms, in the water column as dissolved inorganic phosphorus (usually orthophosphate), as dissolved organic phosphorus, and particulate phosphorus. In sediments, particularly anaerobic sediments, the chemistry of phosphorus can be complicated (Valiela 1984).

The first studies of phosphorus and marine bivalves used the radioactive isotope of phosphorous (^{32}P) to determine uptake and utilization (Ronkin 1950; Pomeroy and Haskin 1954). The earliest evidence that bivalve filter feeders could play an important role in mineral cycling at the ecosystem level came from Kuenzler's (1961) study of the phosphorus budget of the salt marsh mussel in Georgia. Kuenzler concluded that although these bivalves moved about 1% as much energy as bacteria in the salt marsh system, they are probably more important as remineralizers than consumers as a result of their high rates of filtration and deposition.

Only a few studies have focused on the role of oysters in processing phosphorus at the ecosystem level. Sornin et al. (1986) examined the role of intensively cultured *Crassostrea gigas* in the phosphorus cycle in Marennes-Oléron Bay of France. These investigators found that oysters removed about 50% of both total and assimilable phosphorus from the water column of the bay, with most being deposited in the sediments. As earlier described, Dame et al. (1989) determined the phosphorus budget for an intertidal oyster reef, *Crassostrea virginica*, in South Carolina (Table 15). There was a significant uptake of total phosphorus with most of that being particulate. Only 8% of the total phosphorus uptake on the SC reef was released as orthophosphate as compared to 3% by the less dense community in Marennes-Oléron Bay. The expected N:P Redfield ratio is 16, and for the SC reef the ratio for the uptake fluxes is about 2. Consequently, the oyster reef appears to be taking up proportionally more phosphorus than nitrogen or carbon (Dame et al. 1989). Thus, in the oyster-dominated systems studied to date, these animals seem to remove considerably more phosphorus from estuarine waters than would be expected. Oyster reefs may be significant components in retaining phosphorus — a constituent that unlike nitrogen and carbon has no gaseous state and can only enter the estuary via land runoff or import from the coastal ocean.

Asmus et al. (1995) examined the exchange of phosphorus between a mussel bed and the tidal waters of Königshafen bay near Sylt in the Eastern Wadden Sea (see Chapter 5 for detailed description). Using the "Sylt" flume, these investigators concluded that the mussel bed increased the flux of phosphorus both in the release of dissolved inorganic phosphorus and in uptake of particulate phosphorus. They noted that differences in the phosphorus fluxes from bed to bed could be attributed to the environmental characteristics of a given community. Low phosphate release rates from coastal sediments have been attributed to the buffering capacity of the sediments (Pomeroy et al. 1965; Hinga 1990) caused by the adsorption–desorption equilibrium of phosphate interacting with ferrooxihydroxides in interstitial water (Sundby 1992). When the sediments are anoxic, phosphate is released in great amounts due to a stronger reducing environment (Patrick and Khalid 1974; Balzer et al. 1983). Dense mussel bed coverage tends to lead to highly anoxic sediments because of decomposition processes in the organically enriched sediments (Asmus et al. 1995). These same investigators were able to show experimentally that beds with large macroalgae populations released less phosphate than beds with no macroalgae (Asmus et al. 1990). Therefore, young or cultured beds with relatively little macroalgae cycle their phosphorus back into the water column, while mature natural beds store phosphorus in macroalgal biomass.

Silicon

Silicon (Si) and silica (SiO_2) can be limiting nutrients in systems dominated by diatoms. While most animals and algae have a minimal need for silicon, silica forms the ridged cell wall or frustule of diatoms and may account for half of the cell's mass. Most silica is thought to originate from the dissolution of rocks and enters the ocean via rivers. Because it is readily taken up by diatoms and converted into skeletal material, silica sedimentates easily and accumulates in marine deposits. Silica can be recycled by animal metabolism and by release from anoxic sediments. As the demand for silicon by diatoms is high, in those systems dominated by diatoms, it may become a limiting nutrient, particularly in the absence of recycling (Libes 1992).

Silicon can be taken up by bivalves in the process of feeding on and metabolizing diatoms as a food source with the subsequent excretion of silicate or silicic acid (H_4SiO_4) by the animals and the associated sediments (Asmus et al. 1990). Although poorly represented in the literature, there have been a few investigations of bivalves and bivalve beds as sources and sinks of silicon (Asmus 1986; Doering et al. 1987; Dame et al. 1991a; and Prins and Smaal 1994). These studies indicate a relatively high release of silicate from bivalve beds. For example, Prins and Smaal (1994) reported on the estimated uptake of particulate organic silicon and measured the release of silicate by mussel beds in the Oosterschelde estuary (Table 16). Silicate release rates were lower than biogenic silicon uptakes in April and June 1987, when high phytoplankton concentrations caused a high uptake of biogenic silicon. High silicate release was observed in the night tidal cycle of June 1987 and by comparison implies that diatoms are rapidly taken up during the day and remineralized during the evening. High silicate release in the fall probably resulted from the increased rate of dissolution of silicate at higher water temperatures. In addition, concurrent direct measurements of individual bivalve excretion of silicate were below detection limits and suggest that the majority of the silicate release came from the sediments. Thus, it appears that almost all of the biogenic silicon filtered from the water will be deposited and then mineralized in the sediments with return to the water column (Prins and Smaal 1994).

System Nutrient Turnover

The concept of turnover is defined as the ratio of throughput to content (Odum 1983). Turnover is often expressed as a rate fraction or as a unit of time that is the reciprocal of the rate fraction. The turnover rate is the fraction of the total amount

TABLE 16
A Silicon Budget for Mussel Beds in the Oosterschelde Estuary.

Month	Particulate Organic Silica Uptake (mmol m^{-2} h^{-1})	Silicate Release (mmol m^{-2} h^{-1})
April	2.32 ± 0.23 ***	1.11 ± 0.27 ***
June 1987	3.63 ± 0.68 ***	3.20 ± 1.21 **
June 1988	0.56 ± 0.12 ***	0.80 ± 0.50 ns
September	0.15 ± 0.02 ***	3.27 ± 1.45 *

Note: A t-test was used to estimate if the fluxes were significantly different from zero; ns = not significant;

* = $p < 0.05$; ** = $p < 0.01$; *** = $p < 0.001$

After Prins, T.C. and Smaal, A.C. 1994.

of a substance in a system that is released or that enters in a given length of time. Turnover time is the reciprocal of the turnover rate and represents the time required to replace or remove a quantity of material equal to the amount of that material in the system. The residence time is similar to the turnover time and refers to the time a given material remains within the system. With regard to bivalve-dominated systems, a comparison of nutrient turnover times to water mass turnover times can, as we saw in Chapter 5 with grazing, give an indication of the potential for bivalves to regulate the cycling of a particular nutrient.

To date, the concept of nutrient turnover as influenced by bivalves has been applied to ammonium in three ecosystems: North Inlet in South Carolina (Dame et al. 1989), and the Western Wadden Sea and Oosterschelde in the Netherlands (Dame et al. 1991b). In Table 17, the turnover times of ammonium and water residence times are compared for the three systems. Ammonium turnover times were determined from total bivalve uptake and release rates in relation to the average total amount of ammonium within the system. Nutrient turnover times are a function of the biomass of bivalves per unit water volume, the material concentrations in the water column, and other environmental factors including temperature and water velocity. Shorter turnover times relative to water mass turnover time imply a greater influence by the bivalves. The systems represented span the range from a short water mass residence time estuary, North Inlet, to moderate residence time systems in the Netherlands. Ammonium turnover times are about half the water mass residence times in North Inlet and the Oosterschelde, but about double the residence time in the Western Wadden Sea. Thus, it is probable that bivalves are more likely to influence nutrient cycling in the prior systems rather than the latter (Smaal and Prins 1993). These conclusions have been further supported by the field experiments of Dame and Libes (1993) and the modeling efforts of Herman and Scholten (1990).

TABLE 17
Turnover Time of Ammonium Derived from Bivalve Release Rates in Three Different Estuarine Ecosystems and Compared to Water Residence Time in the Same Systems

System	Bivalve Biomass ($g\ m^{-3}$)	Ammonium Release Rate ($gN\ g^{-1}\ h^{-1} \times 10^{-5}$)	Ammonium Turnover Time (Days)	Water Residence Time (Days)
North Inlet	3.8	–3.8	0.5	1
Wadden Sea	3.7	–7.4	18.0	10
Oosterschelde	3.7	–9.5	21.0	40

Coupling Metabolism And Nutrients

Bivalves couple the benthic habitat to the water column and in the process of catabolizing organic compounds not only take up oxygen, but release inorganic nutrients including ammonium and dissolved inorganic phosphate. From the perspective of the animal, the ratio of atoms of oxygen consumed to nitrogen released can be used as an index of a bivalve's rate of catabolism and the composition of its food (Bayne 1976). In the case of nitrogen, the major release product of bivalves is ammonium (see Chapter 3). If the amino acids produced by catabolism are deaminated and the ammonia is excreted while the carbon components of the amino acids are completely oxidized, the theoretical minimum for the ratio of O:N is 7, signifying protein catabolism. Higher values for the O:N ratio indicate increased catabolism of carbohydrate and/or lipid (Bayne 1976; Mayzaud and Conover 1988). As bivalve systems are mostly composed of animals, it is possible that the O:N ratio may have some meaning at this higher level of ecological organization (Dame et al. 1992). Boucher and Boucher-Rodoni (1988) reported an O:N ratio for *Crassostrea gigas* of 21:1 and the surrounding sediments of 34:1. Dame et al. (1992) found that on intertidal *Crassostrea virginica* reefs during the winter the ratio was about 30:1. Both studies suggest that carbohydrate and lipid are the main organic compounds being metabolized. As these oyster reef systems contain animals, plants, microbes, and nonbiological chemical processes, these ratios provide only a relative indication of food utilization (Dame et al. 1992).

CONCLUSIONS

In coastal ecosystems with dense populations of bivalves, these animals remove large quantities of suspended particulate organic materials and remineralize them into forms that are readily utilized by phytoplankton (Figure 38). The remineralization process is amplified in reef systems, in that feces and pseudofeces enrich the sediments surrounding the bivalves, and the microorganisms in these sediments effectively double the remineralization rates to the reef. These processes effectively

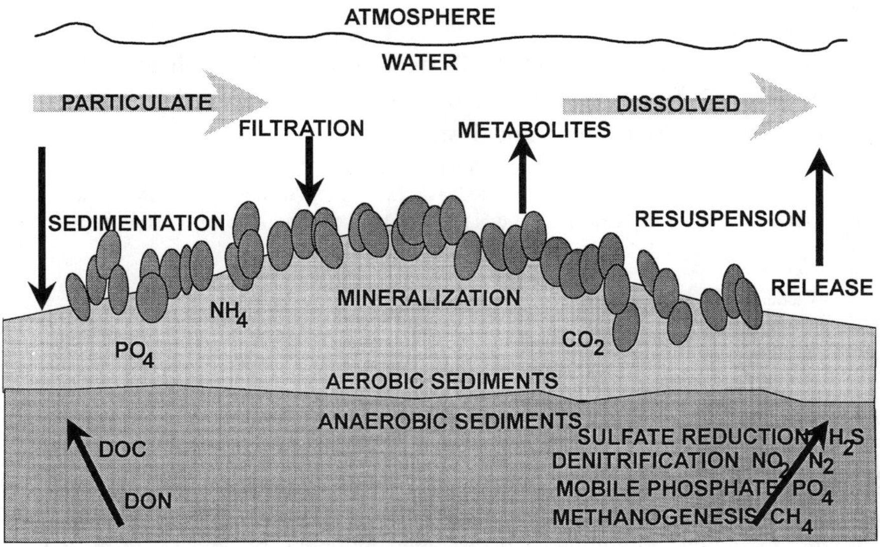

FIGURE 38. A graphical summary of nutrient processes in and around dense systems of bivalves. (Adapted from Dame, R.F. 1993.)

short-circuit the typical pelagic food web and move carbon, nitrogen, and phosphorus through these ecosystems at much faster rates. As a consequence of these material flows, both negative and positive feedback loops are established (Table 18) that potentially increase the productivity and stability of estuarine ecosystems. In essence, dense beds or reefs of bivalves increase the functional and structural sustainability of their ecosystems.

TABLE 18
A Summary of System Nutrient Cycling Control by Oyster Reefs

	Feedback Loop	
Nutrient	(+)	(–)
Carbon	No	Direct
Nitrogen	Direct	Direct
Phosphorus	(no)	Direct

REFERENCES

Alpine, S.E. and Cloern, J.E. 1992. Trophic interactions and direct physical effects control phytoplankton biomass and production in an estuary, *Limnol. Oceanogr.*, 37, 946–955.

Asmus, H., Asmus, R.M., and Reise, K. 1990. Exchange processes in an intertidal mussel bed: a Sylt-flume study in the Wadden Sea, *Ber. Biol. Anst. Helgol.*, 6, 1–79.

Asmus, H., Asmus, R.M., and Zubillaga, G.F. 1995. Do mussel beds intensify the phosphorus exchange between sediment and tidal waters?, *Ophelia*, 41, 37–55.

Asmus, R. 1986. Nutrient flux in short-term enclosures of intertidal sand communities, *Ophelia*, 26, 1–18.

Bahr, L.M. 1976. Energetic aspects of the intertidal oyster reef community at Sapelo Island, Georgia (USA), *Ecology*, 57, 121–131.

Baird, D., McGlade, J.M., and Ulanowicz, R.E. 1991. The comparative ecology of six marine ecosystems, *Phil. Trans. R. Soc. Lond. Ser. B*, 333, 15–29.

Baird, D. and Milne, H. 1981. Energy flow in the Ythan estuary, Aberdeenshire, Scotland, *Estuarine Coastal Shelf Sci.*, 13, 455–472.

Baird, D. and Ulanowicz, R.E. 1989. A seasonal dynamics of the Chesapeake Bay ecosystem, *Ecol. Monogr.*, 59, 329–364.

Balzer, W., Grasshoff, K., Dieckmann, P., Haardt, H., and Petersohn, U. 1983. Redox-turnover at the sediment/water interface studied in a large bell jar system, *Oceanolog. Acta*, 6, 337–344.

Bayne, B.L. 1976. *Marine Mussels: Their Ecology and Physiology,* Cambridge University Press, Cambridge, U.K., 506 pp.

Bertness, M.D. 1984. Ribbed mussels and *Spartina alterniflora* production in a New England salt marsh, *Ecology*, 65, 1794–1807.

Boucher, G. and Boucher-Rodoni, R. 1988. In situ measurement of respiratory metabolism and nitrogen fluxes at the interface of oyster beds, *Mar. Ecol. Prog. Ser.*, 44, 229–238.

Boucher-Rodoni, R. and Boucher, G. 1990. *In situ* study of the effect of oyster biomass on benthic metabolic exchange rates, *Hydrobiologia*, 206, 115–123.

Boynton, W.R., Kemp, W.M., Osborne, C.G., Kaumeyer, K.R., and Jenkins, M.C. 1981. Influence of water circulation rate on in situ measurements of benthic community respiration, *Mar. Biol.*, 65, 185–190.

Cloern, J.E. 1982. Does the benthos control phytoplankton biomass in South San Francisco Bay?, *Mar. Ecol. Prog. Ser.*, 9, 191–202.

Collier, A. 1959. Some observations on the respiration of the American oyster *Crassostrea virginica* (Gmelin), *Publ. Inst. Mar. Sci. Univ. Texas*, 6, 92–108.

Copeland, B.J. and Duffer, W.R. 1964. Use of a clear plastic dome to measure gaseous diffusion rates in natural waters, *Limnol. Oceanogr.*, 9, 494–499.

Dame, R.F. 1993. The role of bivalve filter feeder material fluxes in estuarine ecosystems, in Dame, R.F., Ed., *Bivalve Filter Feeders in Estuarine and Coastal Ecosystem Processes*, Springer-Verlag, Heidelberg, 245–269.

Dame, R.F., Dankers, N., Prins, T. Jongsma, H., and Smaal, A. 1991a. The influence of mussel beds on nutrients in the Western Wadden Sea and Eastern Scheldt estuaries, *Estuaries*, 14, 130–138.

Dame, R.F. and Libes, S.L. 1993. Oyster reefs and nutrient retention in tidal creeks, *J. Exp. Mar. Biol. Ecol.*, 171, 251–258.

Dame, R.F. and Patten, B.C. 1981. Analysis of energy flows in an intertidal oyster reef, *Mar. Ecol. Prog. Ser.*, 5, 115–124.

Dame, R.F., Spurrier, J.D., Williams, T.M., Kjerfve, B., Zingmark, R.G., Wolaver, T.G., Chrzanowski, T.H., McKellar, H.N. and Vernberg, F.J. 1991b. Annual material processing by a salt marsh-estuarine basin in South Carolina, USA, *Mar. Ecol. Prog. Ser.*, 72, 153–166.

Dame, R.F., Spurrier, J.D., and Wolaver, T.G. 1989. Carbon, nitrogen and phosphorus processing by an oyster reef, *Mar. Ecol. Prog. Ser.*, 54, 249–256.

Dame, R.F., Spurrier, J.D., and Zingmark, R.G. 1992. In situ metabolism of an oyster reef, *J. Exp. Mar. Biol. Ecol.*, 164, 147–159.

Dame, R.F., Zingmark, R.G., and Haskin, E. 1984. Oyster reefs as processors of estuarine materials, *J. Exp. Mar. Biol. Ecol.*, 83, 239–247.

Dankers, N., Dame, R.F., and Kersting, K. 1989. The oxygen consumption of mussel beds in the Dutch Wadden Sea, *Scient. Mar.*, 53, 473–476.

DeAngelis, D.L. 1992. *Dynamics of Nutrient Cycling and Food Webs.*, Chapman and Hall, London, 270 pp.

DeAngelis, D.L., Post, W.M., and Travis, C.C. 1986. *Positive Feedback in Natural Systems*, Springer-Verlag, Berlin, 290 pp.

Doering, P.H., Kelly, J.R., Oviatt, C.A., and Sowers, T. 1987. Effect of the hard clam *Mercenaria mercenaria* on benthic fluxes of inorganic nutrients and gases, *Mar. Biol.*, 94, 377–383.

Doering, P.H. Oviatt, C.A., and Kelly, J.R. 1986. The effects of the filter-feeding clam *Mercenaria mercenaria* on carbon cycling in experimental marine mesocosms, *J. Mar. Res.*, 44, 839–861.

Edwards, R.W. and Owens, M. 1964. The oxygen balance of streams, in Goodman, G.J., Edwards, R.W., and Lambert, J.M., Eds., *Ecology and Industrial Society*, Wiley, New York, 149–172.

Feuillet-Girard, M., Heral, M., Sornin, J.-M., Deslous-Paoli, J.-M., Robert, J.-M., Mornet, F., and Razet, D. 1988. Nitrogenous compounds in the water column and at the sediment–water interface in the estuarine bay Marennes-Oléron: influence of oyster farming, *Aquat. Living Resour.*, 1, 251–265.

Hammen, C.S., Miller, H.F., and Geer, W.H. 1966. Nitrogen excretion of *Crassostrea virginica, Comp. Biochem. Physiol.*, 17, 1199–1200.

Hargrave, B.T. and Phillips, G.A. 1981. Annual *in situ* carbon dioxide and oxygen flux across a subtidal marine sediment, *Estuarine Coastal Shelf. Sci.*, 12, 725–737.

Herman, P.M.J. and Scholten, H. 1990. Can suspension-feeders stabilise estuarine ecosystems?, in Barnes, M. and Gibson, R.N., Eds., *Trophic Relationships in the Marine Environment*, Aberdeen University Press, Aberdeen, 104–116.

Hinga, K.R. 1990. Alteration of phosphorus dynamics during experimental eutrophication of enclosed marine ecosystems, *Mar. Poll. Bull.*, 21, 275–280.

Hopkinson, C.S. 1985. Shallow-water benthic and pelagic metabolism: evidence of heterotrophy in the nearshore Georgia Bight, *Mar. Biol.*, 87, 19–32.

Jordan, T.E. and Valiela, I. 1982. A nitrogen budget of the ribbed mussel, *Geukensia demissa*, and its significance in nitrogen flow in a New England salt marsh, *Limnol. Oceanogr.*, 27, 75–90.

Kaspar, H.F., Gillespie, P.A., Boyer, I.C., and MacKenzie, A.L. 1985. Effects of mussel aquaculture on the nitrogen cycle and benthic communities in Kenepuru Sound, Marlborough Sounds, New Zealand, *Mar. Biol.*, 85, 127–136.

Kautsky, N. and Evans, S. 1987. Role of biodeposition by *Mytilus edulis* in the circulation of matter and nutrients in a Baltic coastal ecosystem, *Mar. Ecol. Prog. Ser.*, 38, 201–212.

Kautsky, N. and Wallentinus, I. 1980. Nutrient release from a Baltic *Mytilus*-red algae community and its role in benthic and pelagic productivity, *Ophelia*, (Suppl.), 1, 17–30.

Kemp, W.M., and Boynton, W.R. 1980. Influence of biological and physical processes on dissolved oxygen dynamics in an estuarine system: implications for measurement of community metabolism, *Estuarine Coastal Mar. Sci.*, 11, 407–431.

Kenney, B.E., Litaker, W., Duke, C.S., and Ramus, J. 1988. Community oxygen metabolism in a shallow tidal estuary, *Estuarine Coastal Shelf Sci.*, 27, 33–43.

Kitchell, J.F., O'Neill, R.V., Webb, D., Gallepp, G.W., Bartell, S.M., Koonce, J.F., and Ausmus, B.S. 1979. Consumer regulation of nutrient cycling, *Bioscience*, 29, 28–34.
Kuenzler, E.J. 1961. Phosphorus budget of a mussel population in a Georgia salt marsh, *Limnol. Oceanogr.*, 6, 400–415.
Libes, S.M. 1992. *An Introduction to Marine Biogeochemistry*, John Wiley, New York, 734 pp.
Liebig, J. 1840. *Chemistry in Its Application to Agriculture and Physiology*, Taylor and Walton, London, 401 pp.
Maestrini, S.Y., Robert, J.-M., Lefley, J.W., and Collos, Y. 1986. Ammonium thresholds for simultaneous uptake of ammonium and nitrate by oyster-pond algae, *J. Exp. Mar. Biol. Ecol.*, 102, 75–98.
Manahan, D.T., Wright, S.H., Stephens, G.C., and Rice, M.A. 1982. Transport of dissolved amino acids by the mussel *Mytilus edulis*: demonstration of net uptake from natural seawater, *Science*, 215, 1253–1255.
Marzolf, E.R., Mulholland, P.J., and Steinman, A.D. 1994. Improvements to the diurnal upstream–downstream dissolved oxygen change technique for determining whole-stream metabolism in small streams, *Can. J. Fish. Aquat. Sci.*, 51, 1591–1599.
Mayzaud, P. and Conover, R. 1988. O:N atomic ratio as a tool to describe zooplankton metabolism, *Mar. Ecol. Prog. Ser.*, 45, 289–302.
McCarthy, J.J. 1980. Nitrogen, in Morris, I., Ed., *The Physiological Ecology of Phytoplankton*, University of California Press, Berkeley, 191–234.
Murphy, R.C. and Kremer, J.N. 1985. Bivalve contribution to benthic metabolism in a California lagoon, *Estuaries*, 8, 330–341.
Newell, R.C. and Field, J.G. 1983. The contribution of bacteria and detritus to carbon and nitrogen flow in a benthic community, *Mar. Biol. Lett.*, 4, 23–36.
Nixon, S.W., Oviatt, C.A., Garber, J., and Lee, V. 1976. Diel metabolism and nutrient dynamics in a salt marsh embayment, *Ecology*, 57, 740–750.
Nixon, S.W., Oviatt, C.A., Rogers, C., and Taylor, K. 1971. Mass and metabolism of a mussel bed, *Oecologia*, 8, 21–30.
Nixon, S.W. and Pilson, M.E.Q. 1983. Nitrogen in estuarine and coastal marine ecosystems, in Carpenter, E.J. and Capone, D.G., Eds., *Nitrogen in the Marine Environment*, Academic Press, New York, 565–648.
Odum, E.P. 1983. *Basic Ecology*, Saunders, Philadelphia, 613 pp.
Odum, H.T. 1956. Primary production in flowing waters, *Limnol. Oceanogr.*, 1, 102–117.
Odum, H.T. and Hoskin, C.M. 1958. Comparative studies of the metabolism of marine waters, *Publ. Inst. Mar. Sci. Texas*, 5, 16–48.
Patrick, W.H and Khalid, R.A. 1974. Phosphate release and sorption by soils and sediments: effect of aerobic and anaerobic conditions, *Science*, 186, 53–56.
Pomeroy, L.R. 1970. The strategy of mineral cycling, *Annu. Rev. Ecol. Syst.*, 1, 171–190.
Pomeroy, L.R. and Haskin, H.H. 1954. The uptake and utilization of phosphate ions from sea water by the American oyster, *Crassostrea virginica* (Gmel.), *Biol. Bull.*, 107, 123–129.
Pomeroy, L.R., Smith, E.E., and Grant, C.M. 1965. The exchange of phosphate between estuarine water and sediments, *Limnol. Oceanogr.*, 10, 167–172.
Prins, T.C. and Smaal, A.C. 1994. The role of the blue mussel *Mytilus edulis* in the cycling of nutrients in the Oosterschelde estuary (The Netherlands), in Nienhuis, P.H. and Smaal, A.C., Eds., *The Oosterschelde Estuary (The Netherlands): A Case Study of a Changing Ecosystem*, Kluwer, Dordrecht, 413–429.
Robert, J.M., Maestrini, S.Y., Heral, M., and Zanette, Y. 1982. Production des micro-algues des claires ostreicoles en relation avec l'azote ornique dissous excrete par les huitres, *Oceanol. Acta*, 1989, 389–395.

Rodhouse, P.G., Ottway, B., and Burnell, G.M. 1981. Bivalve production and food chain efficiency in an experimental nursery system, *J. Mar. Biol. Assoc. U.K.*, 61, 243–256.

Rodhouse, P.G. and Roden, C.M. 1987. Carbon budget for a coastal inlet in relation to intensive cultivation of suspension-feeding bivalve molluscs, *Mar. Ecol. Prog. Ser.*, 36, 225–236.

Ronkin, R.R. 1950. The uptake of radioactive phosphate by the excised gill of the mussel, *Mytilus edulis*, *J. Cell. Comp. Physiol.*, 35, 241–260.

Smaal, A.C. and Prins, T.C. 1993. The uptake of organic matter and the release of inorganic nutrients by bivalve suspension feeder beds, in Dame, R.F., Ed., *Bivalve Filter Feeders in Estuarine and Coastal Ecosystem Processes*, Springer-Verlag, Heidelberg, 273–298.

Sornin, J.-M., Feuillet, M., Héral, M., and Fardeau, J.-C. 1986. Influence des cultures d'hitres *Crassostrea gigas* sur le cycle du phosphore en zone intertidale: role de la biodeposition, *Oceanol. Acta*, 9, 313–322.

Sprent, J.I. 1987. *The Ecology of the Nitrogen Cycle*, Cambridge University Press, New York, 151 pp.

Stephens, G.C. 1982. Recent progress in the study of 'Die Ernährung der Wassertiere und der Stoffhaushalt der Gewässer, *Am. Zool.*, 22, 611–619.

Sundby, B., Gobeil, C., Silberberg, N., and Mucci, A. 1992. The phosphorus cycle in coastal marine sediments, *Limnol. Oceanogr.*, 37, 1129–1145.

Ulanowicz, R.E. and Tuttle, J.H. 1992. The trophic consequences of oyster stock rehabilitation in Chesapeake Bay, *Estuaries*, 15, 298–306.

Underwood, A.J. 1994. On beyond BACI: sampling designs that might reliably detect environmental disturbances, *Ecol. Appl.*, 4, 3–15.

Valiela, I. 1984. *Marine Ecological Processes*, Springer-Verlag, New York, 546 pp.

Warwick, R.M., Joint, I.R. and Radford, P.J. 1979. Secondary production of the benthos in an estuarine environment, in Jefferies, R.L. and Davy, A.J., Eds., *Ecological Processes in Coastal Environments*, Blackwell Scientific, Oxford, 429–450.

Wright, R.T., Coffin, R.B., Ersing, C.P., and Pearson, D. 1982. Field and laboratory measurements of bivalve filtration of natural marine bacterioplankton, *Limnol. Oceanogr.*, 27, 91–98.

Wright, S.H. 1982. A nutritional role for amino acid transport in filter feeding marine invertebrates, *Am. Zool.*, 22, 621–634.

7 Ecosystem Experiments: Models, Experimental Ecosystems, and Field Manipulations

INTRODUCTION

Ecosystems in which bivalves play important roles are complex. Bacterial and decompositional components adjacent to bivalves may operate only with anaerobic conditions. Phytoplankton grazed for food may be carried long distances by the currents while functioning on very much shorter time scales than the bivalves. Finally, the bivalves themselves may play significant roles as major benthic–pelagic coupling mechanisms. To address ecosystem scale questions, ecologists have traditionally utilized a number of experimental approaches (Table 19) that incorporate increasing system complexity. These approaches are system level computer simulation models, living models or micro- and mesocosms, and full-scale ecosystem field manipulations (Table 19). Thus, there is a scaling up from simplifications of the real system that depends on the purpose of the exercise and is representative of the state of knowledge at the conception of the model to direct observations of real systems and their undisturbed behavior.

Historically, most ecological studies consisted of making observations and then conceiving mentally probable explanations for the observations. Hairston (1989) calls these explanations or conclusions "weak inference," because there are frequently alternative explanations, and there is little assurance as to which conclusion is correct. He further argues that manipulative experiments provide a preferable alternative, because planning experiments require at least an implied prediction. Ecosystem level computer simulation models usually offer more scientific rigor than direct interpretations of observations. The development of mathematical hypothesis to explain processes and mechanisms in specific systems can result in predictions and force their makers and users to think more clearly. Unfortunately, such models may only explain the obvious and, more important, incorporate assumptions that are difficult to justify (Hairston 1989). If, however, these models are well conceived and validated, they have the potential to provide an experimental platform for systems too large and complex to directly manipulate (Herman and Scholten 1990).

Intermediate to the more or less completely controlled mathematically based ecosystem models and minimally controlled field experiments are the laboratory or

TABLE 19
Experimental Approaches Used in Bivalve Research, with Relative Estimates of Temporal and Spatial Scales, and Ecological Complexity

Approach	Organizational Level	Temporal Scale	Spatial Scale	Processes	Ecological Complexity
Endoscope	Organism	Seconds to minutes	mm	Filtration	Low
Jar	Organism	Minutes to hours	cm	Respiration Filtration Excretion	Low
Microcosm	Organisms	Hours	cm	Respiration Filtration Excretion	Moderate
Exclusion cages	Populations to communities	Months to years	m	Predation Competition	High
Lab flume	Population	Hours	m	Grazing	Moderate
Field flume and tunnels	Community	Hours	m	Grazing Nutrients Metabolism	Moderate to high
Simulation models	Populations to ecosystems	Days to years	m to km	Multiple	Moderate
Mesocosm	Simple system	Days to year	m	Multiple	Moderate +
Incidental experiments	Ecosystem	Months to years	km	Multiple	High
Designed experiments	Ecosystem	Months to years	km	Multiple	High

Adapted from Parker and Newell unpublished.

field chambers also known as micro- or mesocosms. In these experimental ecosystems, most environmental variables are controlled or at least vary in a systematic manner, and a single variable is manipulated to address a specific question. In addition, these chambers are replicated to show repeatability and provide statistical rigor. Because the chambers themselves are smaller and less dynamic than a real ecosystem, there are often complications and unrealistic findings. Other potential problems include pseudoreplication, edge effects, changing controls, and stratification (Hairston 1989).

Unlike the preceding approaches to ecological experiments, with field manipulation experiments, it is not possible to exercise control over most of the variables in the system. In this approach, the variable of interest is manipulated in a set way while the other variables fluctuate independent of the experiment (Hairston 1989). Ecosystem level field experiments have come into great favor in recent years because many of the important features of ecosystems are at large temporal and spatial scales. In estuaries with dense communities of bivalves, for example, water dynamics, biogeochemical processes, and predation are difficult to reproduce in chambers and not sufficiently understood to describe in mathematical simulation models. Often these field experiments are not easily replicated because of spatial and temporal constraints and, ultimately, costs (Carpenter et al. 1995).

In this chapter, each category of ecosystem experiment, models, experimental chambers, and field manipulations will be addressed with particular attention to those systems dominated by marine bivalves. With regard to bivalve-dominated ecosystems, these approaches are not self-exclusionary and there is often a blending of two or more of these techniques in the search for knowledge.

MODELS

Models are simplified versions of the real world. In science as elsewhere, informal models may be graphical as is a diagram, picture, or a painting or verbal as a description of a process or cycle. These types of models are often useful to present a concept in its most basic form. Many informal ecological models are transformed into formal models with statistical or mathematical representations that allow quantitative predictions of the processes and states within the modeled system. These models formally relate ecological interactions, both biological and physical, through specific mathematical relationships. In addition to simulation modeling, steady-state (budget) models are often analyzed using the engineering–economic techniques of flow and network analyses. This latter approach provides a systems level view of material cycling, flow structure and connections, and system development. It also allows the direct comparison of very different ecosystems through the use of unitless indices.

Computer simulations of models have become a normal part of the modeling process. A computer simulation model allows the researcher to make predictions of potential outcomes of the model and to experiment with the various parameters and components of the model in ways that would be inappropriate in a natural system. Also, the development of the model usually provides a summary or accounting of what is known about the system of interest and provides the ecologist with information on areas needing new or better information.

The development of models often begins with the construction of a conceptual diagram, flow chart, or box model, many of which have been used throughout this book. At this level of development, there are four basic components of the model: (1) forcing functions or properties outside of the system of note that drive, regulate, or control the system; (2) state variables or components that describe what we see in the model, i.e., populations, trophic levels, etc.; (3) flows of materials or energy between the various components often depicted as arrows; and (4) interaction functions that formalize how interactions modify, amplify, or control flows (Odum 1983). Every good model will describe the space or boundaries of the system, the important subsystems in the larger system, and the time interval of interest in the model. Once a simulation model is built, it must be validated by comparing the model outcomes to independent observations of the real system. If these observations do not confirm the simulation of the model, then investigator must reexamine or reevaluate the model for the validity of the relationships or assumptions applied or for the appropriateness of the data utilized. The modeling process is a clear application of the scientific method. For detailed information on ecosystem model building, readers are directed to Patten (1971), Hall and Day (1977), Halfon (1979), Platt et al. (1981), Odum (1983), Ulanowicz and Platt (1985), and Hannon and Ruth (1994).

There are many more models that depict various aspects of bivalves and their relationships to the environment than space permits for review. Some of these models are emerging, e.g., Grant et al. (1993 and 1995), Emerson et al. (1994), and Hatcher et al. (1994), who are developing physiological based carrying capacity models linked to water flow dynamics, and some are controversial, e.g., Powell et al. (1992, 1994), and Hofman et al. (1992, 1994, 1995), who have developed a budget model for oysters that is linked to a detailed hydrodynamic numerical model. In the remaining portions of this section, several bivalve-dominated and ecosystem level simulation models will be presented, and two different steady-state budget models will be examined from a systems analytical perspective.

System Simulation Models

Bivalve Carrying Capacity of Marennes-Oléron

As described in Chapter 5, the Bay of Marennes-Oléron, France, is one of the highest bivalve producing areas in Europe. Examination of over a century of data on oyster production in Marennes-Oléron (Héral et al. 1986) indicated that there were three distinct phases. The low production period prior to 1927 was related to the development of Portuguese oyster (*Crassostrea angulata*) culture and the exploitation of natural oyster beds. Between 1927 and 1960, production increased four-fold and this was attributed to increased recruitment through spat collectors and improved breeding techniques. From 1960 to 1969, overexploitation caused a dramatic decline in growth rates and production. The high densities of oysters resulted in animals of poor condition that led to an epidemic that resulted in the loss of *C. angulata* from the system. Since 1969, *C. angulata* has been replaced by the Japanese oyster, *Crassostrea gigas*, and previous bivalve production levels have been reestablished.

In a series of papers (Bacher 1989; Bacher et al. 1991; Héral 1993; Raillard et al. 1993), the investigators of this bay describe a group of models they have developed to estimate bivalve-carrying capacity by Marennes-Oléron.

The Marennes-Oléron group first developed population models based on the assumption that, for a given time frame, environmental factors have a constant mean (Héral 1993). Using their long time series of historical data on production, growth, and mortality rates for the two species of oysters, they were able to compute standing stock sizes comparable to those actually observed. The establishment of a relationship between biomass or standing stock showed an asymptotic value of about 40,000 tons per year production for the bay for both species of oysters. The observed data fit the von Bertalanffy relationship well (Chapter 4) and the asymptotic value defines the carrying capacity of the bay for bivalves (Héral 1993). In addition to showing the carrying capacity for the system, this investigation utilizes biomass and production instead of the normal population units of numbers and incorporates the physical environment in the initial estimates. These strategies allowed this group to make the transition to their next model. This model is a more typical ecosystem representation of carbon flow in the Marennes-Oléron system.

After acquiring an extensive data set on the major components in the Marennes-Oléron system and in order to evaluate the importance of these components for future research effort, Héral (1993) constructed a carbon flow model (conceptual) for this ecosystem (Figure 39). The model included the fluxes of particulate and dissolved carbon, phytoplankton, zooplankton, benthos, oysters, phytobenthos, and sediments. Using this accounting approach, Héral (1993) concluded that the zoo-benthos and zooplankton were on a relative basis not as important as the other components.

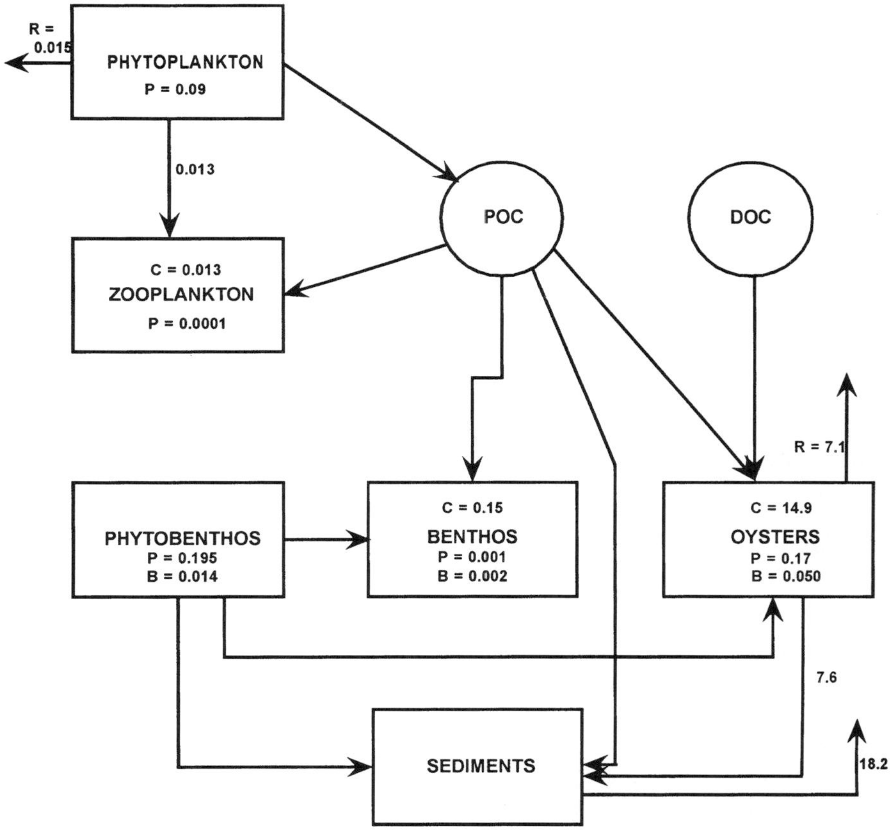

FIGURE 39. Conceptual model of the carbon budget for Bay of Marennes-Oléron, France. (Adapted from Héral 1993.)

These investigators developed a group of models that in total described the main trophic interactions in the bay. Included in this group was a numerical hydrodynamic model that divided the oyster production study area (Figure 40) into 13 boxes (Bacher 1989). The size of a box is of the same order of magnitude as the movement of an average water particle by tidal action on a single tide. The water volume of each

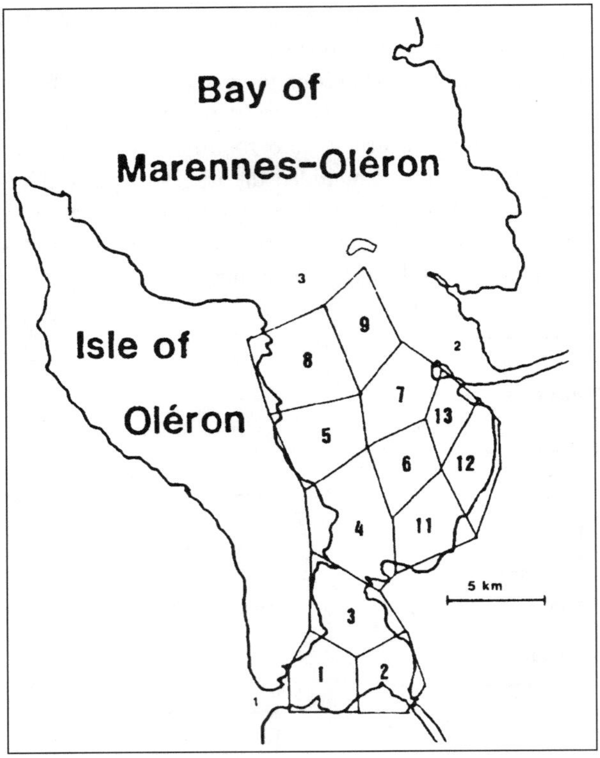

FIGURE 40. Location and grid for Marennes-Oléron numerical model for water flow and oyster population simulation. (Adapted from Bacher 1989.)

compartment was driven by tides and wind, and transport between boxes was governed by the conservation of mass relationship. The transport of nutrients and suspended particulate material was driven by the hydrodynamic model and coupled to the three main biological models for phytoplankton, phytobenthos and oysters (Bacher 1989; Bacher et al. 1991; Héral 1993). The oyster model (Raillard et al. 1993) is based on the fundamental energy budget relationships expressed earlier in Chapters 3 and 4. The only state variable in the oyster model is the soft body biomass or flesh weight. This variable is controlled by the inputs and outputs of such physiological processes as filtration rate, pseudofeces production, ingestion, oxygen consumption as a measure of metabolic rate, and spawning. The model is designed to incorporate the effects of the quantity and quality of food, temperature, and oyster weight. The exact formulations of the major components of the energy budget for *C. gigas* are given in Tables 20a and 20b and the details of the sources of the given relationships are provided by Raillard et al. (1993).

One of the more difficult tasks in developing and utilizing this type of overall model is the coupling of the various submodels. Time and spatial scales for both the hydrodynamic and physical parameters must be consistent with the biological models. By utilizing a box or compartment approach (spatial scale) for the oyster

TABLE 20a
Equations Representing the Processes in the Oyster Model for Marennes-Oléron Bay

	Food Consumption
Particulate consumption	$Q_{[d,p,m]} = SES_{[d,p,m]} \cdot F$
Total consumption	$Qt = F/(Ws^{bf}) \cdot SES$
Filtration rate	$F = F0 \cdot e^{(kf \cdot \min(0, Ts - SES))} \cdot Ws^{bf}$
Total seston	$SES = SES_d + SES_p + SES_m$
	Ingestion
Phytoplankton and detritus	$I_{[d,p]} = (1 - PF_0) \cdot Q_{[d,p]}$
Inorganic particulates	$I_{[m]} = (1 - PF_m) \cdot Q_m$
Pseudofeces production	$PF_{[o,m]} = PFX_{[o,m]} \cdot (1 - e^{(kp[o,m] \cdot \min(0,1-Qt))}) + (1 - PFX_{[o,m]}) \cdot (1 - e^{(kp2 \cdot \min(0,C2-Qt))})$
	Absorption
Absorption	$A_{[d,p]} = ae \cdot I_{[d,p]}$
Energy absorbed	$EA_{[d,p]} = A_{[d,p]} \cdot X_{[d,p]}$
Absorption efficiency	$ae = aet \cdot TMP + ae0$
Detritus energy conversion	$X_d = (ES - SES_p \cdot X_p)/SES_d$
Total seston energy	$ES = SES_L \cdot X_L + SES_G \cdot X_G + SES_p \cdot X_p$
	Metabolic Costs
Respiration	$R = (art \cdot TMP + ar0) \cdot Ws^{br}$
Total metabolic costs	$ER = R \cdot X_{O_2}$
	Reproduction
Total spawning effort	$P = ap \cdot Wj^{bp}$
Daily spawning effort	$p = P^{(1/dp)}$
	Scope for Growth
Scope for growth	$DwJ/DT = ((EA_d + EA_p) - ER) \cdot 1 - 10^{-3} \cdot pc \cdot TIM - p \cdot Wj$
Caloric value of oysters	$Ws = Wj \cdot Y_h$

From Héral 1993; Raillard et al. 1993.

production area of Marennes-Oléron Bay, the Lagrangian residuals for water flow between boxes can be calculated in relation to the tides. Thus, the time scale becomes the tidal cycle (Bacher 1989). Resuspension, sedimentation, and material fluxes are calculated for each compartment in the spatial grid. The oyster model is computed for two age classes in each box, and oyster growth is a function of the food made available by the hydrodynamic model. Predators and competitors are specific to each box and operate as driving variables on oyster growth. The model was validated by comparing simulated and observed oyster growth (Bacher et al. 1991). In summary, this model is very bivalve oriented and functions well in its main objective to predict bivalve-carrying capacity in Marennes-Oléron Bay.

TABLE 20b
Definition of Terms in Marennes-Oléron Oyster Model

Parameter	Definition
ae	Absorption efficiency
ae0	Absorption efficiency at t = 0°C
aet	Slope of absorption as a function of temperature
ap	Constant for spawning effort
ar0	Respiration at t = 0°C
art	Slope of respiration as a function of temperature
$A_{[d,p]}$	Particulate absorption
bf	Allometric exponent for filtration
bp	Allometric exponent for spawning
br	Allometric exponent for respiration
C1, C2	Two levels of pseudofeces production
dp	Length of spawning period
$EA_{[d,p]}$	Energy absorbed
ER	Metabolic costs
ES	Total energy of seston
F	Filtration rate
F0	Standard filtration rate
$I_{[d,p,m]}$	Particulate ingestion
kf	Filtration exponent
kp	Pseudofeces production exponents
p	Daily ratio of weight for spawning effort
P	Total ratio of the weight for the spawning effort
pc	Percentage of absorbed energy in soft body growth
$PFX_{[o,m]}$	Threshold of pseudofeces production
$PF_{[o,m]}$	Rate of pseudofeces production
Qt	Total consumption
$Q_{[p,d,m]}$	Particulate consumption
R	Respiration
$SES_{[p,d,m,P,L,G]}$	Seston concentration in water column
TIM	Fraction of time oysters immersed
TMP	Water temperature
Ts	Level of clogging
W_j	Caloric value for oyster bodies
W_s	Dry weight of oyster bodies
X_{O_2}	Energy conversion for oxygen
X_p	Energy conversion for chlorophyll
$X_{[d,P,L,G]}$	Energy conversions for seston
Y_h	Weight conversion for the caloric value of oyster bodies
Y_p	Weight conversion for chlorophyll

Note: Subscripts d, p, m, o, P, L, and G refer, respectively, to detrital organic matter, phytoplankton, particulate inorganic matter, particulate organic matter, particulate proteins, particulate lipids, and particulate carbohydrates.
Source: Héral 1993; Raillard et al. 1993.

SMOES: Simulation Model Oosterschelde Ecosystem

Since 1979, the Oosterschelde estuary (Eastern Scheldt) of the Netherlands has been extensively modified by the building of a storm-surge barrier in the mouth of the estuary and two large auxiliary compartment dams in the upper portions of the system. As a consequence of this engineering marvel, the regional human population has been protected from North Sea storms, and because of reduced tidal exchange and river input the estuary now behaves as tidal bay. The system is characterized by high production rates, tight benthic-pelagic coupling, and a high degree of biological control between components. Because the engineering changes to the Oosterschelde system would result in profound changes in the extant habitats and many of the management options taken influence the complex interactions by which the components of the aquatic ecosystem depend on each other, simulation modeling was chosen as an essential tool for predicting the consequences of these management choices on the structure and function of the ecosystem (Herman 1994). The goal of the SMOES model is to provide a quantitative summary of the scientific observations and to provide a tool for analysis and prediction of the ecosystem's responses to the engineered manipulations (Klepper et al. 1994). SMOES describes the main carbon and nutrient flows with a spatial scale of 10 to 20 km and a temporal scale of approximately one day. For a more detailed documentation of SMOES, see Klepper (1989).

MODEL FORMULATION

SMOES simulates 11 state variables: diatoms, other algae, zooplankton, silicate, inorganic nitrogen, silicon detritus, carbon detritus, benthic diatoms, salinity, refractory detritus, and oxygen. The food web for the system is given in Figure 41, note that filter feeders are a forcing function and not a state variable (Klepper 1989). Spatially the ecosystem is divided into four boxes, each with a tidal excursion of 10 to 15 km (Figure 42). Model structure is the same for each box, but morphological parameters such as depth, surface area, etc., are compartment dependent. The simulation consists of formulating the rate of change in every state variable as a result of the state of the system itself and of time-dependent external conditions, e.g., light intensity, temperature, nutrient fluxes, and state variable concentrations in the adjacent North Sea. After calculating all the rates of change for a given time increment, the new state of the system is calculated and then moves ahead one time step (Klepper et al. 1994)

The water transport submodel (Klepper et al. 1994) makes the assumptions that: (1) the bivalve filter feeders have a fixed position in the moving water mass and (2) these animals only feed from the compartment in which they reside at midtide. The first assumption has been examined in detail by Klepper (1989) and found to be acceptable. The second is well within the spatial resolution of the model. As dispersive transport dominates advective transport in the Oosterschelde, the transport of dissolved materials was modeled according to a "forward time centered space" approximation to the advection–dispersion model (Ruardij and Baretta 1989) and are described in detail by Klepper et al. (1994). Particulate transport is modeled

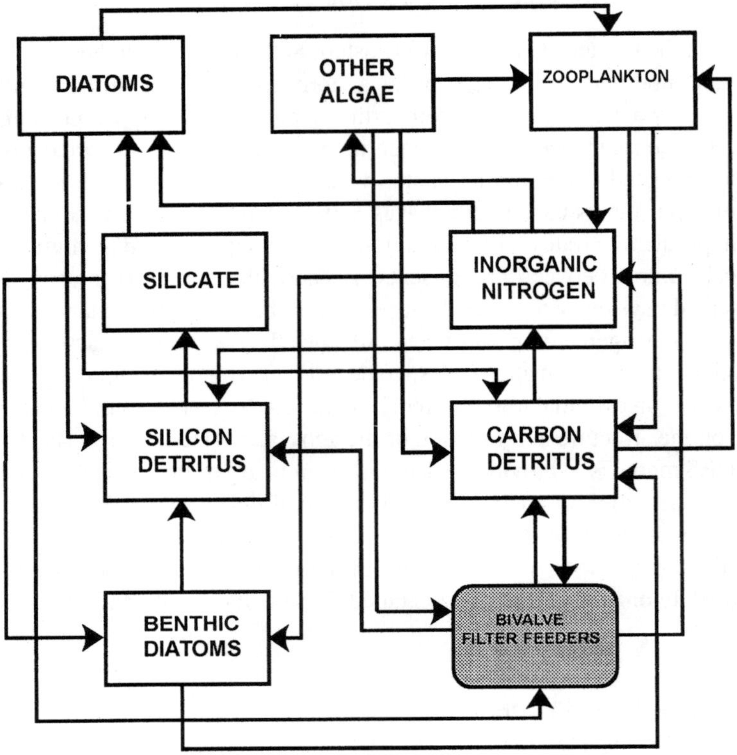

FIGURE 41. Conceptual model of the SMOES simulation of the Oosterschelde estuary, The Netherlands. (Redrawn from Klepper 1989.)

similarly to that of dissolved substances, but account is made for particles that settle out of suspension or are resuspended during a tidal cycle (Klepper et al. 1994)

The major biological submodels are those of Phytoplankton, Microphytobenthos, Zooplankton, Macrobenthos and Mineralization. The details of these submodels are presented in Klepper (1989) and Klepper et al. (1994) and their formulations are described in Tables 21a and 21b.

The phytoplankton model (Tables 21a and 21b) is based on the work of Eilers and Peeters (1988 and 1993). It describes photosynthesis and primary production as a function of light intensity and dynamic processes. There are two unique aspects to the phytoplankton model. First, photosynthesis is based on a physiological model and is assumed to take place in phytoplanktonic units in three states: resting, active and inactive. Second, there are two different phytoplankton groups in the Oosterschelde: diatoms and nondiatoms. Each of these aspects is taken into account in the modeling process. The microphytobenthos is modeled in a similar fashion (Klepper 1989).

The zooplankton are defined as both holoplankton and meroplanktonic larvae, and the model takes both into account. The main objective of this component is its

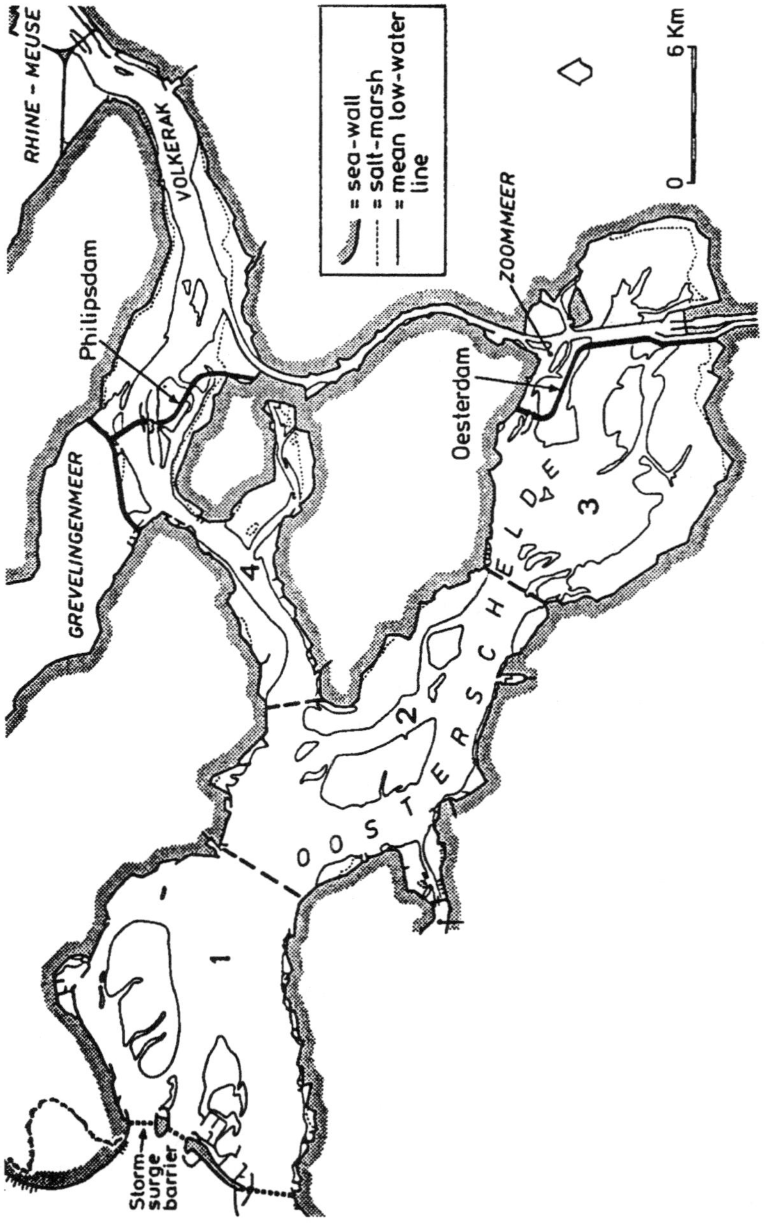

FIGURE 42. SMOES locational map showing location of the simulation grid. (Redrawn from Klepper 1989.)

TABLE 21a
Equations Representing the Biological Submodels in the SMOES Simulation

Phytoplankton and Microphytobenthos	
Photosynthesis	$P = \frac{I}{aI^2 + bI + c}$
Integrated gross production	$p_g = D_l \cdot P_{max} \cdot F_l$
Maximum production (temp)	$P_{maxt} = P_{max10}\exp((T - 10)T_{coef})$
Nutrient limitation	$F_n(C) = \frac{C}{Ks + C}$
Chlorophyll content	$Chl = \frac{D_l \cdot F_n}{1.34I_0}$
Chlorophyll:carbon	$Chl: C = Chl_{max} F_n (1 - F_l)^{0.6}$
Respiration	$R_p = R_o + (1 - pv) \cdot (P_g - R_o)$
Excretion	$E = E_{max}(1 - F_n)P_g$
Zooplankton	
Ingestion	$R_t = R_{max\,t} \min\left\{1, \frac{F}{F_{lim}}\right\}$
Mortality	$Z_{mort} = mqqB$
Macrobenthos	
Deposit feeder ingestion	$G = c_T B_g(B_p + B_d)$
Filter feeder clearance	$CR = f_T g_s aW^b$
Assimilation efficiency	$AE = AE_{max}POM_q/(POM_q + 0.15)$
Respiration	$R_b = f_T pW^q$
Pseudofeces	$PSF = CR_{max}\{0,(S - S_{psf})\}$
Nutrient Processes	
Mineralization	$M = k(T) \cdot DET$
Denitrification	$DC = NC \cdot C \cdot r_{10} \cdot f(T)\left[\frac{N}{K_m + N}\right]$

influence in grazing the phytoplankton and follows the approaches of DiToro et al. (1971) and Kremer and Nixon (1978).

The macrobenthos is composed of two groups: the deposit feeders (grazing bottom algae and detritus) and the suspension feeders (feeding on detritus and phytoplankton). The deposit feeders are dominated by meiofauna (nematodes and harpacticoid copepods), surface deposit feeders (the snail *Hydrobia*) and infaunal deposit feeders (particularly the worm *Arenicola*). The dominant suspension feeders are cockles (*Cerastoderma*), mussels (*Mytilus*), and fouling organisms on the rocky substrate introduced in the engineering process. As the mussels are all cultivated and the cockels are heavily managed as a fishery, the biomass of these groups is used to empirically drive the model (Klepper et al. 1994).

TABLE 21b
Definitions of Parameters Used in SMOES Simulation Equations

Parameter	Definition
a	Coefficient
AE	Assimilation efficiency of macrobenthos
AE_{max}	Maximum assimilation efficiency of macrobenthos
b	Coefficient
B	Biomass of zooplankton
B_d	Bentic labile detritus
B_g	Biomass of benthic deposit feeder
B_p	Bomass of phytobenthos
c	Coefficient
C	Bottom labile detritus concentration
Chl	Chlorophyll *a* content
Chl:C	Chlorophyll to carbon ratio
Chl_{max}	Maximum chlorophyll *a* content
CR	Clearance rate of filter feeders
CR_{max}	Maximum clearance rate of filter feeders
C_T	Deposit feeder temperature-dependent ingestion coefficient
D_l	Day length
Dc	Denitrification rate in carbon units
DET	Detritus concentration
E	Phytoplankton excretion rate
E_{max}	Maximal phytoplankton excretion as a fraction of gross production
F	Zooplankton food concentration
F_l	Phytoplankton dimensionless reduction function
F_{lim}	Zooplankton threshold food concentration
F_n	Phytoplankton nutrient limitation function
f_T	Macrobenthic exponential temperature function
G	Deposit feeder ingestion rate
g_s	Negative exponential function of suspended sediments
I	Light intensity
I_o	Light intensity at water surface
K_m	Denitrification half-saturation coefficient
K_s	Nutrient limitation half saturation concentration
k(T)	Mineralization rate as a function of temperature
M	Mineralization
mqq	Proportionality constant for zooplankton mortality
N	Inorganic nitrogen concentration
NC	Nitrogen consumption per unit of carbon consumption
P	Photosynthesis
p	Respiration coefficient
P_g	Daily integrated gross primary production per unit volume per unit biomass
P_{max}	Maximum primary production
P_{maxt}	Maximum primary production as a function of temperature
P_{max10}	Maximum primary production at 10°C

TABLE 21b (continued)
Definitions of Parameters Used in SMOES Simulation Equations

Parameter	Definition
pv	Production value, the amount of biomass produced per unit of primary production
POM_q	Fraction of organic matter in macrobenthos food
PSF	Pseudofeces production
q	Respiration exponential coefficient
R	Macrobenthic respiration
R_{maxt}	Zooplankton temperature dependent maximum daily ration
R_o	Phytoplankton maintenance respiration
R_p	Phytoplankton respiration
R_t	Zooplankton temperature dependent ingestion per unit biomass
r_{10}	Relative denitrification rate at 10°C and saturated dissolved inorganic nitrogen
S	Suspended matter concentration
S_{psf}	Threshold concentration for pseudofeces production
T	Temperature in °C
T_{coef}	Temperature coefficient
W	Filter feeder dry body weight

Two nutrient processes are modeled, the decomposition of detritus or mineralization and the process of denitrification in the sediments by the microphytobenthos.

MODEL ANALYSIS

The comparison between actual observed data and the simulated output of the SMOES model is described in the analysis of uncertainty (Klepper et al. 1994) and the calibration efforts of Scholten and van der Tol (1994). Inconsistencies between the model and the observed data can result from several causes (Herman 1994):

1. The model formulations may be a poor representation of reality.
2. A number of model parameters are assumed to be constant and are in reality variable.
3. Model inputs are only approximately known and lead to model uncertainty.

After analyzing SMOES, Klepper et al. (1994) concluded that the weakest parameters or parts of the model were processes that had not been well studied in the project. These weaknesses included: microbiological processes, algal respiration, zooplankton food limitation and loss processes, carbon-to-chlorophyll ratio, transport of algae and detrital silicon, assimilation efficiency of cockles, and gas exchange with the atmosphere.

The issue of model validation is addressed in detail by Scholten and van der Tol (1994). These workers use the Schelsinger et al. (1979) definition for validation:

substantiation that a computer simulation model within its domain of use possesses a satisfactory range of accuracy consistent with the intended application of the model. More generally stated by Herman (1994), a model is valid if it accurately predicts system behavior under the scenarios for which it was constructed. In order to validate SMOES, this group compared the adequacy of the model simulation runs (are all field data falling within the model's uncertainty limits?) and the reliability of these simulations (how good is the average fit between the model and the data?) (Herman 1994). Depending on the rigor with which one holds to these definitions of validity, SMOES is as valid or invalid as most other ecosystem level models (Scholten and van der Tol 1994).

With particular interest to the major bivalve groups in the Oosterschelde, Herman and Scholten (1990) used the pre-storm surge barrier version of SMOES to examine the relationship between the suspension feeding bivalves and the phytoplankton. By varying the biomass of the bivalves and the input of nutrients to the ecosystem, they were able to investigate the influence of these changes on phytoplankton biomass. They found that bivalve filter feeders could stabilize the phytoplankton under changing nutrient inputs. Filter feeder grazing (Chapter 5) regulated the phytoplankton at high nutrient levels and nutrient concentrations regulated at low levels. At low nutrient levels, the recycling of nutrients (Chapter 5) via the filter feeder grazing was essential to maintaining phytoplankton primary production. These authors also found that although these results are consistent with the field data and post-storm-surge barrier simulations, they cannot be obtained using the pre-barrier version of the model to generate a post-barrier prediction. The differences between real observations and the simulated predictions may be the result of compensating errors within the models (Herman 1994). Regardless of its failings, the SMOES model is the most complicated and extensively analyzed simulation model in which bivalves play a major role at the ecosystem level.

System Analytical Models

Oyster Reef Model

In Chapter 6, the flow of carbon through an oyster reef was introduced, and this conceptualization can also be rendered as an energy flow model for the oyster reef community (Dame and Patten 1981). In addition to providing an accounting of the standing crops and flows of energy in the oyster reef, these authors applied a number of analytical techniques to the model that provided further insights into the processes operating within this community. These techniques are from the realm of systems analysis and are derived from the input–output theory of economics as applied to ecosystems (Hannon 1973). This approach has the advantage of revealing the structure of the system by demonstrating with energy or matter flows the direct and indirect dependence of each state variable or component upon the others. Hannon's initial adaptations have been expanded in succeeding years by Finn (1976), Patten et al. (1976), and Ulanowicz (1986) into a set of formal theories on the analysis of flows within ecosystems. The oyster reef model will be used as an introduction to flow analysis in a system dominated by bivalves.

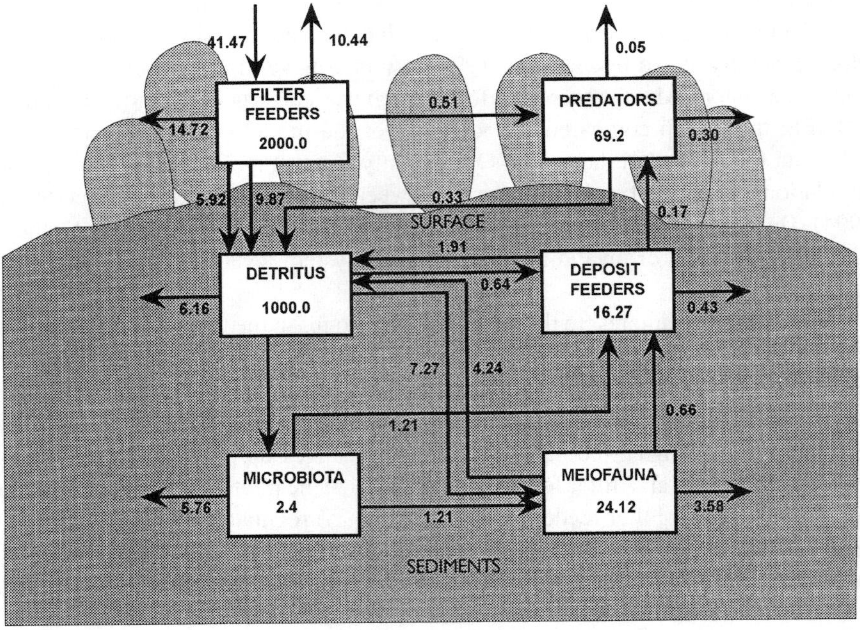

FIGURE 43. Oyster reef energy flow model. Flows are in kcal m^{-2} d^{-1} and states are in kcal m^{-2}. (Data from Dame and Patten 1981.)

Energy flow in the oyster reef system is shown in Figure 43. Detailed definitions for the standing crops (boxes) in Kcal m^{-2} and flows (arrows) in kcal m^{-2} d^{-1} are given in the same publication. In the oyster reef system, causality is transmitted by conservative energy flows and at times by nonconservative information flows or at least flows that are similar to information controls. Input-output flow analysis permits the tracing of relationships within the system. Thus, the use of this analysis allows the investigator to describe the internal flow structure of the system. This structure can show how much energy is flowing through the system vs. how much is cycling within the system. In this particular case, energy cycling means that some potential energy is not utilized on its first pass through a given component and can be used on a subsequent pass through the system. In flow analysis, the initial procedure is to normalize in turn one unit of output from each component in the reef system. Back calculation from an output determines the source flows necessary to produce a given unit of output. These source flows are specific for each unit of outflow and are often called input environments (Figure 44). This figure gives the input flow environment necessary to generate one unit of outflow from the predators in the oyster reef model. The values within the compartments are the throughflows required to generate the unit of outflow; numbers associated with arrows represent causal flows such as feeding or respiration. The sum of the input or output flows to a given component defines the throughflow value (T_{ii}).

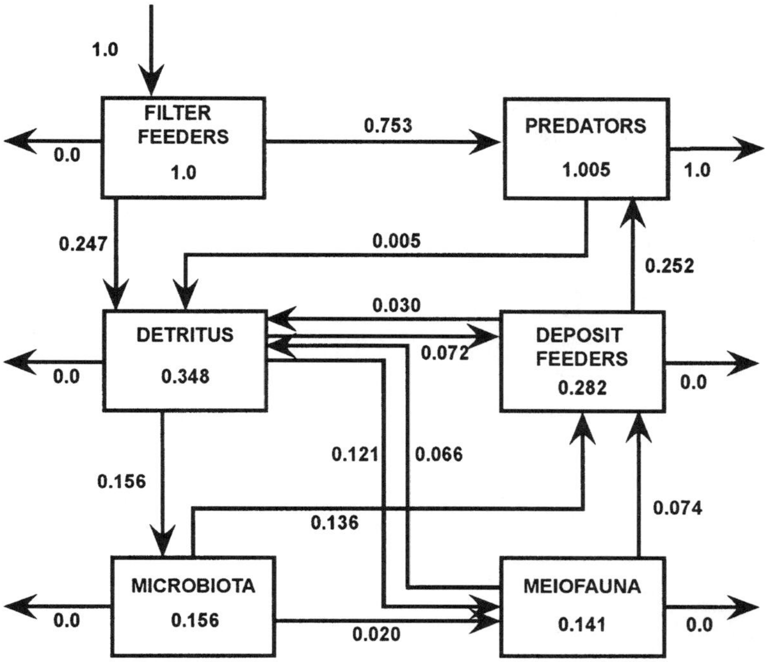

FIGURE 44. Oyster reef model input environments (see text for explanation).

An example of output environments for the oyster reef model is described in Figure 45. In this case, one unit of input into the filter feeders generates the throughflows and outputs shown (Figure 45). The case shown is the only real output environment for this model as there is only one inflow of energy into the system and that is to the filter feeders. Imaginary output environments can be computed for the other components, and these are necessary to calculate other flow analysis measures (Dame and Patten 1981).

At steady-state there are a number of characteristics of flow analysis. They are pathlength, cycling efficiency, and cycling index. Each of these measures is useful in comparing the structure and functional traits of different ecosystems.

Total system throughflow (TST) is the sum of the unnormalized throughflows, and for the oyster model this is 85.58 kcal m^{-2} d^{-1}. Because the oyster reef model conforms to laws of thermodynamics, throughflow observed for each component declines as energy moves through the system (Table 22).

Average pathlength of an inflow (APLZ) is defined as the average number of compartments through which a given inflow passes (Finn 1976). Thus, for the oyster reef model, 1 unit of energy enters the filter feeders and has a pathlength of 1.0 (Figure 42). Only 39% of this input moves on to other components, while the remaining 61% leaves the system as respiration and egestion. Of that energy that moves on, 39% will have a pathlength of at least 2.0 and 24% will have a pathlength

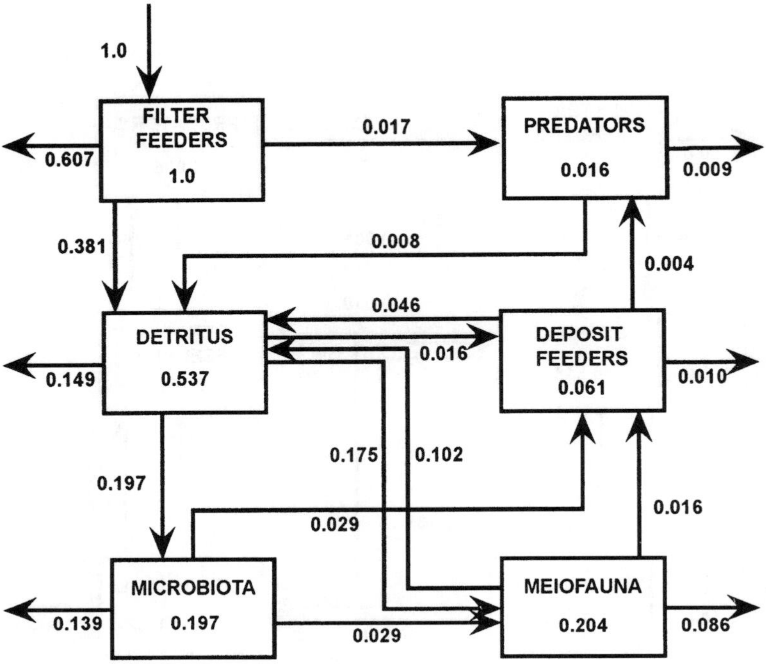

FIGURE 45. Oyster reef model output environments (see text for explanation).

TABLE 22
Flow Analysis Measures for the Oyster Reef Model

Measure	Filter Feeders	Detritus	Microbiota	Meiofauna	Deposit Feeders	Predators
Throughflow	41.47	22.27	8.17	8.48	2.51	0.69
Average pathlength of unit input	2.02	2.59	1.84	2.54	3.12	2.23
Average pathlength of unit output	1.00	2.93	3.93	4.07	4.71	2.93
% Total system throughflow as inflow	100.00	—	—	—	—	—
% Total system throughflow as outflow	30.10	21.66	27.10	17.45	2.43	1.26
Recycling efficiency	0	0.28	0.09	0.23	0.11	0.01

of 3.0. The result is an average pathlength of an inflow of 2.02. The average pathlength of a unit of outflow (APLY) from a given compartment is derived similarly to APLZ (Table 22).

In the oyster reef model, all energy does not move directly or straight through the system; some energy cycles. Energy cycling does not mean energy is degraded more than once, but that some potential energy is not utilized on its first pass through a component and a small proportion of that energy may be recycled back to a component it has already passed through. Energy cycling is particularly common in the sediments where deposit feeding organisms are constantly ingesting and egesting partially decomposed material. Thus, in the oyster reef model, cycling pathways are depicted among the deposited detritus, microbiota, meiofauna, deposit feeders, and predators (Figure 43).

A recycling efficiency (RE) may be calculated for a given component by subtracting 1.0 unit of flow from the throughflow for that particular component in the outflow environment for that component ($T_{ii} - 1.0$) and dividing by the same throughflow (T_{ii}). The REs for the oyster reef model are given in Table 22.

From the preceding, it is implied that total system throughflow (TST) is composed of a cycled (TST_c) and a noncycled or straight (TST_s) portion. By multiplying the appropriate recycling efficiency (RE) in Table 22 by the throughflow for a given component, the portion of throughflow due to cycling can be calculated. Summing the cycling portions yields total system throughflow which cycles (TST_c) and by subtraction, the energy that moves directly through the system (TST_s). In the case of the oyster reef model, total system throughflow that cycles is 9.21 kcal m^{-2} d^{-1}. A dimensionless cycling index (CI) may also be calculated by dividing cycled throughflow (TST_c) by total system throughflow (TST). For the oyster reef system, this index is 0.11 (Dame and Patten 1981). Therefore, 89% of the energy entering the system flows directly through the system and 11% is recycled and most of this is in the sediment portions of the model.

The analysis of the flow structure of the oyster reef model shows the separation of the filter feeder compartment from the rest of the system because there are no feedback energy flows to the filter feeders from other components in the system. Visual analysis of the structure of the flows in the sediment portion of the model reveals numerous feedback loops that can influence the processes there (Dame and Patten 1981). This portion of the model is where 11% of the energy is recycled and is supported by the elevated (relative) RE for the detritus, microbiota, meiofauna, and deposit feeder components.

Marsh Fertilization and Bivalve Production

Flow analysis was used by Finn and Leschine (1980) to investigate the proposition that by fertilizing *Spartina alterniflora* salt marshes, more food would be produced for bivalve filter feeders, and thus bivalve production would be increased. Salt marsh fertilization studies in both Great Sippewissett marsh in Massachusetts (Valiela et al. 1975) and Barataria Bay, Louisiana (DeLaune et al. 1976) show that both systems can significantly increase the yield of *S. alterniflora* through the addition of nitrogen fertilizers. The bivalve of interest in Great Sippewissett is *Modiolus* (= *Geukensia*) *demissa* and in Barataria Bay *M. demissa* and *Crassostrea virginica*. Existing models of nitrogen flow were utilized (Leschine 1979 and Day et al. 1973) for the respective systems.

Using only the models stated above, Finn and Leschine (1980) were able to deduce through flow analysis that only a small portion of the nitrogen added to salt marsh sediments finds its way into the bivalves — 3% of added nitrogen in the

Sippewissett model and 8% in the Barataria Bay model. In these salt marsh systems, most of the nitrogen flows through *S. alterniflora* that is strongly coupled to the sediments where the nitrogen is applied. As noted by Dame and Patten (1981), the filter feeding bivalves are not strongly coupled to the sediments, but to the water column. Thus, flow analysis can be used to examine questions concerning the management of systems where direct information is lacking.

Chesapeake Bay Ecosystem Model and Oysters

The initial carbon flow model of the Chesapeake Bay ecosystem was developed by Baird and Ulanowicz (1989). This model focused on the carbon budgets and exchanges of the 36 most important components in the mesohaline portion of the bay and examined these interactions in a seasonal perspective. An aggregated 13-component version of this model is depicted in Chapter 6 (after Ulanowicz and Tuttle 1992).

Ulanowicz and his colleagues have subjected the Chesapeake Bay carbon model to flow analysis, using similar measures to those utilized on the oyster reef model and have developed new and unique network (flow) analysis tools that further elucidate the structure of this system (Ulanowicz 1986; Baird and Ulanowicz 1989; Baird et al. 1991; Wulff and Ulanowicz 1989). One of these tools is very graphical, the Lindeman Spine. In this analysis the energy flow networks are mapped into simplified trophic aggregations and depicted with their associated routes of recycling. Trophic efficiencies are also given as percentages within the boxes and are defined as that fraction of the total carbon in a trophic level that is transmitted to the next higher trophic level. Respiration and exports from each level are also shown (Figure 46). In the 36-compartment version of the Chesapeake Bay carbon model, the trophic chain is 8 levels long, and oysters reside within level II. Using an aggregated 15-compartment version of the Chesapeake Bay carbon model, Baird et al. (1991) calculated an average pathlength (APLY) of 3.61 and cycling index (CI) of 29.7.

During the 1980s, Ulanowicz (1980, 1986) developed additional measures of a system's flow or network analysis that he argued showed the degree of network development. As discussed by Baird et al. (1991), the measures were an elaboration of specific information theory concepts. Ulanowicz (1986) contended that the amount of species richness and the extent of trophic specificity (the lack of trophic niche and overlap) are embodied in the average mutual information of the flow connections. To have high network mutual information, a system should have many species or functional groups of comparable size that are unambiguously connected with each other. That is, each component exchanges matter or energy with only a very few other compartments. Ulanowicz (1980) scaled the mutual information of the network of flows by its activity or total system throughflow (TST), and named the resulting measure system ascendancy (A). Ascendancy represents the combined attributes of system size and trophic organization (see Ulanowicz 1986 for a detailed development of this theory).

Briefly, Ulanowicz (1986) quantifies the diversity of pathways in a flow network (food web) as follows:

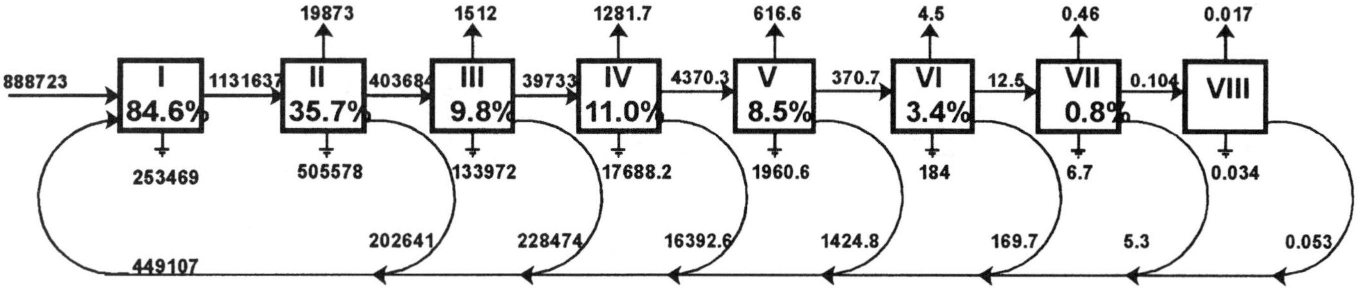

FIGURE 46. The trophic chain or Lindeman spine for the Chesapeake Bay carbon model; flows in mgC m^{-2} $year^{-1}$. (Adapted from Baird and Ulanowicz 1989.)

$$C = -k \sum_i Q_i \log Q_i$$

where C is the capacity of the system and a quantification of the diversity of pathways in the flow network, Q_i is the ratio of the quantity of material or energy flowing through a given compartment to the total system throughflow, and k is an arbitrary constant. Note the similarity of this relationship to the Shannon-Weaver information index for species diversity:

$$H = -k \sum_i p_i \log p_i$$

Because C does not quantify the size or relationship between compartments, k has been equated to total system throughflow (T_i)and the probability of flow between any two components (f_{ij}). The result is a measure described by Ulanowicz (1986) as network ascendancy (A) and formulated as:

$$A = T \sum_k \sum_j f_{kj} Q_k \log \left[f_{kj} \Big/ \left(\sum_i f_{ij} Q_i \right) \right]$$

Network ascendancy (A) has both size and network properties, and capacity or system development capacity (C) is the upper limit of (A). Thus, the degree of system development is the fraction of possible organization that is actually realized, i.e., A:C. The actual values for ascendancy, capacity, and their ratio are relative to comparisons with other systems, but the current situation in Chesapeake Bay produces values intermediate to other systems studied (Baird et al. 1991). From their analyses of the Chesapeake Bay model, Baird and Ulanowicz (1989) concluded that under the present circumstances oysters are not central players in the trophic dynamics of this system.

However, Newell (1988) raised the proposition that oysters once were and can once again become key agents in the structuring of Chesapeake Bay's ecosystem. Accordingly, Ulanowicz and Tuttle (1992) aggregated the original 36-compartment model of Baird and Ulanowicz (1989) into a 13-compartment carbon balance model as shown in Chapter 6, Figure 35. This dynamic model renders the current circumstances in Chesapeake Bay and was designed to explicitly address the trophic consequences of increasing oyster stocks (biomass). Ulanowicz and Tuttle (1992) reasoned that oysters in the bay had been overharvested and that by reducing the exploitation rate of oysters only food availability to the oysters and not oyster larval recruitment would be influenced. The results of this simulation of the Chesapeake Bay carbon model under the preceding conditions are shown in Figure 47. The flows and states represented by Figure 47 are for a 23% reduction in the exploitation rate

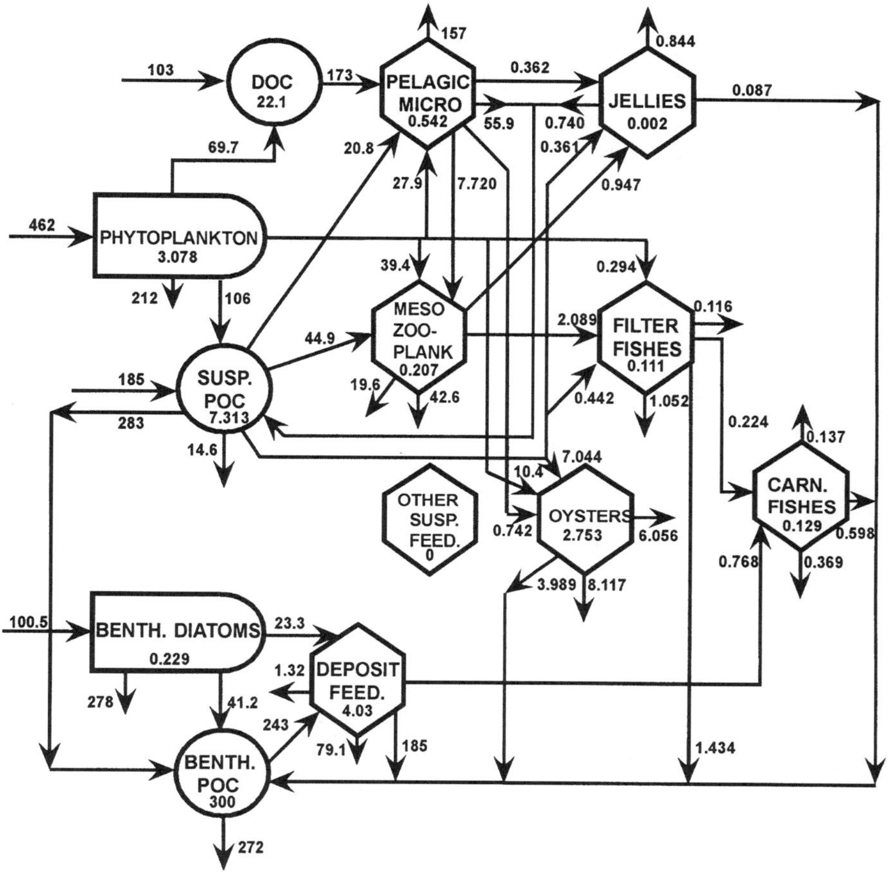

FIGURE 47. The Chesapeake Bay carbon simulation model assuming reduced oyster stock exploitation. (Adapted from Ulanowicz and Tuttle 1992). Flows in gC m^{-2} $year^{-1}$ and states in gC m^{-2}.

of oysters. The most obvious result of this change was an increase in oyster biomass of 150% over their original values (Chapter 6, Figure 35). The extinction of the "other benthic suspension feeders" compartment is considered to be the result of a false assumption concerning competition. A comparison of the current situation with the hypothesized oyster rehabilitation case is given in Table 23. The model predicts that increasing oyster abundances will decrease phytoplankton productivity as well as stocks of pelagic microbes, ctenophores, medusae, and particulate organic carbon. Field data from the vicinity of oyster raft aquaculture plots in the same area of Chesapeake Bay support the model's predictions of phytoplankton productivity, bacterioplankton, and labile organic carbon. Ulanowicz and Tuttle (1992) argue that restoring oyster beds or introducing raft culture represents a potentially significant addition to the goal of mitigating eutrophication through reduction of nutrient inputs.

TABLE 23
Percentage Changes in the Standing Crops of the Compartments in Figure 35 That Are Introduced by a 23% Reduction in the Rate of Oyster Exploitation per Unit Biomass as Shown in Figure 46

Compartment (State)	Percentage Change
Increases	
Oysters	+150.2
Benthic diatoms	+29.1
Carnivorous fishes	+17.5
Filter feeding fishes	+5.2
Mesozooplankton	+4.8
DOC	+1.3
Decreases	
Gelatinous zooplankton	–89.2
Phytoplankton	–11.5
Pelagic microbes	–6.2
Suspended POC and bacteria	–5.2
Sediment POC and bacteria	–1.8

After Ulanowicz and Tuttle (1992).

Experimental Ecosystems

The Scientific Committee on Ocean Research (SCOR) has defined experimental ecosystems as physically confined, self-maintaining, multitrophic systems with a duration time exceeding the generation time of the penultimate trophic level present, and of a size sufficient to enable pertinent sampling and measurements to be made without seriously influencing the structure and dynamics of the system (Lalli 1990). In the case of bivalves, self-maintaining means that their food is generated within the experimental ecosystem itself, in contrast to an aquarium where food is provided from outside the system (Pilson 1990). From a practical standpoint, these systems are also classified according to size as:

Microcosms	$<1\ m^3$
Mesocosms	$1\ m^3$ to $10^3\ m^3$
Macrocosms	$>10^3\ m^3$

The temporal and spatial requirements of these systems are tightly coupled with larger volumes or areas required for larger organisms or processes that require longer time spans or generation times (Figure 48). The use of such experimental ecosystems is firmly linked to the natural environment, laboratory work, and models. In contrast to computer simulation models, these mini-ecosystems are often referred to as living models and find great utility in studying systems and processes that are difficult to observe in nature or where there is a need to examine human influences on systems without endangering the natural system, i.e., pollution and global change questions.

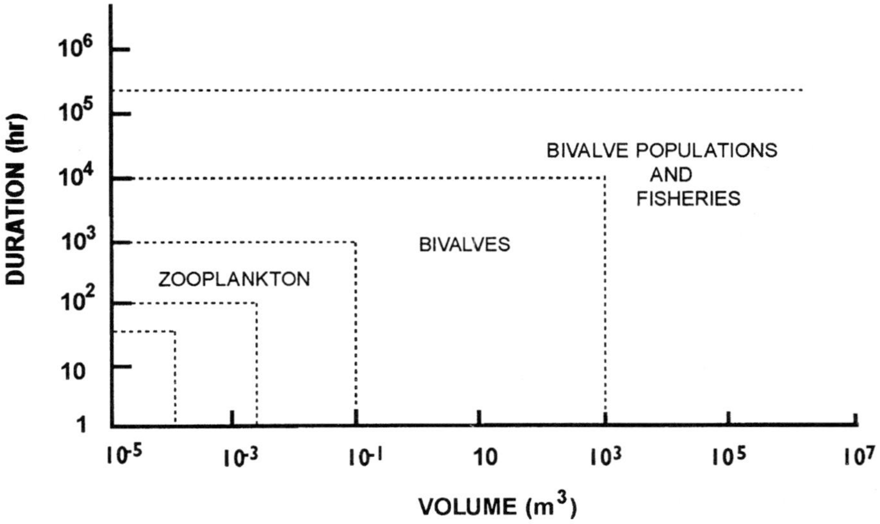

FIGURE 48. Spatial and temporal scales for experimental ecosystems emphasizing bivalves.

Although numerous bivalve studies have used small chambers to measure various parameters, these chambers are not self-sustaining and will not be considered separately in this work. In this section, the role of experimental ecosystems, mesocosms, in ecosystem level questions involving bivalves will be examined. In particular, the boundary conditions concerning the use of these systems will be reviewed and their unique utility in answering specific ecosystem level questions will be examined.

MERL, Rhode Island, U.S.A.

The MERL, Marine Ecosystem Research Laboratory, system of mesocosms is adjacent to Narragansett Bay, Rhode Island. As the focus of mesocosm-based research since the 1970s, the main objectives of the MERL are to obtain insight into the biological interactions of complex systems; to investigate the effects of pollutants on ecosystems; and to study the biogeochemistry of selected pollutants in the estuarine environment (Pilson et al. 1980). The MERL tanks are land-based stirred chambers with a benthic sediment component scaled to water volume at the same ratio as Narragansett Bay. Water is exchanged between the tanks and the bay at the same level of magnitude as water exchanges in the adjacent natural system. Extensive initial studies showed that the water within the mesocosms closely resembled that in the Bay with respect to temperature, mixing (Nixon et al. 1980), primary production (Oviatt et al. 1981), nutrient concentrations, and dynamics (Pilson et al. 1980), phytoplankton (Vargo et al. 1982), and benthic community structure (Grassle et al. 1981). A number of preliminary comparison tests showed that the inter- and intratank replicability of these mesocosms was sufficient and adequate to allow extensive experimentation (Oviatt et al. 1981; Smith et al. 1982).

Early in the use of MERL, benthic filter feeding by bivalves in the sediments at the bottom of the chambers was identified as an important process in regulating the removal rates of suspended particles in these mesocosms (de Wilde 1990). This aspect of estuarine processes was examined in detail by the MERL researchers, using the presence or absence of the clam *Mercenaria mercenaria* in a series of experiments of varying duration (Doering et al. 1986, 1987, 1989; Doering and Oviatt 1986). This group's initial experiments utilized four chambers over a 119-day late spring through summer period and used radioactive ^{14}C as a tracer and measure of several rate processes. They found that with regard to carbon cycling (Chapter 6) the presence of clams increased system production, net and gross sedimentation of particulate carbon, and the benthic remineralization of dissolved inorganic nitrogen. The higher production in mesocosms with clams was attributed in part to the greater flux of dissolved inorganic nitrogen from the benthos. Even though clam tanks had greater production, water column biomass was the same in both control and experimental tanks. It was assumed that the clams regulated, but did not limit through filtration, the phytoplankton biomass in the water column (Doering et al. 1986, 1987; Doering and Oviatt 1986; also see Chapter 5).

In another series of experiments (Doering et al. 1989) involving 12 tanks over 365 days, the roles of eutrophication, silicon, and the benthos including bivalve filter feeders were examined. Focusing only on the benthic filter feeder results, these authors found that this functional component was almost as important in explaining the variability of water column plankton as external nutrient loading. In tanks with benthos, the primary grazing control on diatoms was by the benthic filter feeders, while in tanks without benthos, zooplankton appeared to be more important. There was also evidence that the benthic filter feeders altered the community structure of the plankton. The removal of the benthos appeared to allow the planktonic food chains to truncate to higher trophic levels (note current Chesapeake Bay, Chapter 5) and these investigators speculate that by suppressing higher pelagic trophic levels the benthic component will increase its share of available resources as either pelagic meroplankton or as adults in the sediments. The results of these MERL experiments have significantly contributed to the understanding of many benthic–pelagic coupling relationships in coastal waters that would have been difficult to examine otherwise (de Wilde 1990).

Roskilde Fjord, Denmark

Using a highly replicated series of plastic enclosures (Table 24), Riemann et al. (1988, 1990) and Horsted et al. (1988) reported a series of seasonal studies (spring, summer, and autumn) on the regulation of estuarine phytoplankton and zooplankton by suspension feeding bivalves, planktivorous fish, nutrients, and nutrient exchange with the sediments. The 24 unstirred experimental chambers were fixed to a pontoon bridge about 200 m from shore. Eight of the chambers were in contact with the sediments and the remainder were closed. Although there are several major bivalve filter feeder populations in Roskilde Fjord, the mussel, *Mytilus edulis*, was used as

TABLE 24
A Comparison of Experimental Ecosystems, Mesocosms, That Have a Significant Bivalve Component

Name	Location	Type	Number	Vol. (m^3)	Duration (h)	Bivalve Component	Grazing Rate ($m^3\ d^{-1}$)	Nutrient Loading	Ref.
MERL	Rhode Island	Mixed tank	4–12	13	119–8,750	Natural Sediments	1–2	None	Doering et al. 1986, 1989; Doering and Oviatt 1986
Roskilde	Denmark	Plastic bags	24	6.2	17–23	Cages	7	Replacement	Riemann et al. 1988; Horsted et al. 1988
Oosterschelde	Netherlands	Mixed tank	4	3	28	Separate chambers	0.5–1	Replacement	Prins et al. 1995
Oresund	Sweden	Plastic cylinders	8	0.3	12	Suspended net	0.24	Heavy	Olsson et al. 1992; Graneli et al. 1993

the representative filter feeder in the experimental chambers. Using allometric relationships and estimated filtration rates to determine the numbers of mussels needed to represent the natural system, three cages of mussels were suspended at 1, 2, and 3 m depths within the chambers.

Mesocosms with mussels always showed reduced phytoplankton biomass. Mussel growth was low in chambers with only mussels, indicating over-grazing by these bivalves. In chambers with mussels, fish, and nutrients, mussel growth rates were higher. Also, in chambers with nutrient additions, mussels, probably because of their part in inorganic nutrient recycling, reduced the effects of nutrient additions on phytoplankton production. *Mytilus* reduced the number of small zooplankton, tintinnid ciliates, and rotifers, but did not appear to influence the abundance of larger zooplankton species. The larger zooplankton, *Acartia tonsa* and *Pleopis polyphemoides*, were controlled by the planktivorous fish. Thus, these studies offer additional support for bivalve filter feeder regulation of phytoplankton through both nutrient recycling and grazing. They also provide evidence of bivalves regulating smaller zooplankton and protozoans.

Oresund and Skagerrak, Sweden

A Swedish group (Olsson et al. 1992 and Graneli et al. 1993) conducted a series of experiments using land-based mesocosms to investigate the role of consumers (grazers) and nutrients in structuring the phytoplankton community in the Oresund and Skagerrak areas of coastal Sweden. Eight nonstirred plastic cylinders were filled with natural phytoplankton communities and *Mytilus*, zooplankton, zooplankton plus scyphozoan medusae, and fish were added to different chambers experimentally (Table 24). The number of mussels added to the experimental chambers was calculated to be equivalent to the filtration rates of copepods at ambient concentrations. Nutrient concentrations were maintained at a set level. The experiments were conducted during the late spring at Oresund and during autumn at Skagerrak.

Although the phytoplankton communities varied at the two sites, *Mytilus* was found to be capable of grazing all phytoplankton species except monads <3 μm at both sites. Adding higher trophic levels in the form of jellyfish did change phytoplankton community composition, but bivalves did not. Bivalve grazing, however, could lead to an increase in picoplankton abundance (Olsson et al. 1992; Graneli et al. 1993).

Oosterschelde, The Netherlands

As part of the continuing investigation of ecosystem processes in the coastal and estuarine waters of the Netherlands, a set of four land-based mesocosms was established in the early 1990s. These mesocosms (Figure 49) are similar in construction to the MERL chambers, but are smaller (Table 24). The chambers are stirred and have a sediment component. Each tank has a pair of benthic chambers that are designed to contain filter feeders like mussels and enable the quantification of fluxes of particulate and dissolved materials between the bivalve and the water column containers (Prins et al. 1995). After a series of pilot studies, initial experiments were conducted to examine the relationships between mussel grazing and phytoplankton biomass and to study the effects of nutrient recycling by the bivalves on primary and secondary production. This was attempted by trying to stimulate the development of a phytoplankton bloom through the continuous external supply of nutrients and grazing by mussels at four different densities. The experiment took place in late spring.

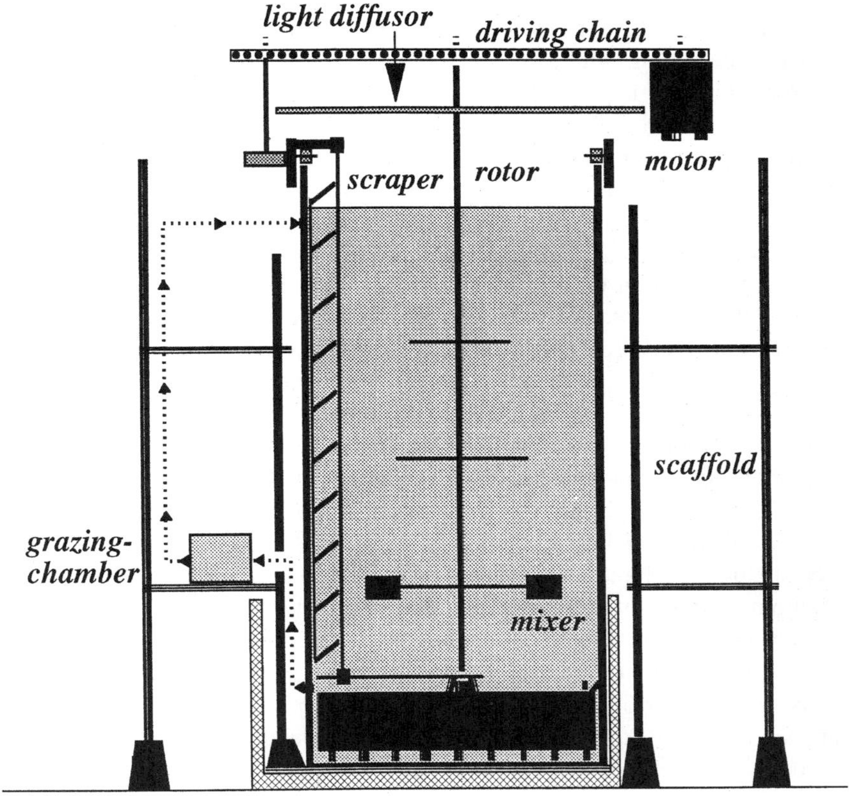

FIGURE 49. A diagram of an Oosterschelde experimental mesocosm.

Phytoplankton biomass was significantly reduced in mesocosms with the highest mussel biomass. Phytoplankton in these chambers had a higher proportion of diatoms than communities in the other mesocosms. Phytoplankton growth rates were highest in mesocosms with high mussel biomass and seemed to imply a shift to the faster growing diatoms and an increase in nutrient availability. Because the reduction in phytoplankton biomass by grazing was higher than the increase of phytoplankton growth rates, total primary production was lowest in high mussel mesocosms. Prins et al. (1995) concluded that under conditions of high external nutrient load, mussel grazing may exert an effective regulation of phytoplankton biomass and may significantly alter process rates and structure in the pelagic system.

Mesocosms and Bivalves: Conclusions

The mesocosm studies described above and summarized in Table 24 are each somewhat different in their particulars. The amount of bivalve grazing in these systems ranges from 0.24 to 7 m^3 d^{-1}. The low value from Olsson et al. (1992) and Graneli et al. (1993) is probably an underestimate (Prins personal communication) and the upper value from Riemann et al. (1988) appears to be overgrazing in that system. In any case, the Danish experiment appears to represent unnatural conditions because the bivalve densities used are unsustainable over a reasonable length of time. Prins (personal communication) has suggested that the shift to picoplankton in this system is typical of conditions with extremely high grazing pressure (rapid water turnover and growth rates of larger planktonic cells smaller than grazing mortality), and typical of short-term experiments where the bivalves are not sustainable due to starvation. Other major differences in these experiments may be the differences in external nutrient loading, ranging from none in the MERL to major in the Danish experiments.

From the observations of four different mesocosm systems (Table 24), evidently this approach has been used in relation to bivalve filter feeders in an attempt to understand the role these animals play in ecosystem processes. Because these studies were carried out under controlled conditions, the dual nature of bivalve filter feeders exerting grazing control on the plankton and nutrient enrichment through remineralization and excretion have been clearly demonstrated. Under the specified conditions of nutrient loading and phytoplankton doubling, bivalves can regulate through grazing intensity both phytoplankton and zooplankton biomass and biodiversity. By taking nutrient remineralization by the adjacent sediments and nutrient excretion by the bivalves themselves into account, it is also clear that these animals can enhance phytoplankton growth rates and primary production. Thus, both top-down (grazing), and bottom-up (nutrient enrichment) control of the phytoplankton can be mediated by the bivalve filter feeders. These studies also support the notion that bivalves can play a significant role in the cycling of nutrients in shallow marine and estuarine habitats (see Chapter 6). However, because mesocosms are only simplifications of nature and only as good as their scaling and validation allow, final substantiation of the processes studied in them remains in the realm of natural ecosystem level manipulations.

Ecosystem Field Experiments

Ecosystem scale experiments on real systems are the most realistic, yet the most difficult to conduct with the specific control of environmental parameters. For bivalves, these experiments have their genesis in the use of bell jars or chambers in the field (Zeitschel and Davies 1978), domes (Boucher and Boucher-Rodoni 1988), and in tunnels (Dame et al. 1984) to elicit direct measurements of material fluxes between the bivalves and their community and the water column. From a different perspective, population and community ecologists have used various devices to exclude predators or competitors from portions of benthic communities dominated by bivalves in order to investigate the roles of predation and competition in community structure (Connell 1972, 1975; Menge 1976; Paine 1977). While these approaches have allowed the estimation of processes at the level of a population or community, they are not able to definitively answer questions on an ecosystem scale and, as they are discussed in previous chapters, they will not be examined further. Instead, unintentional or incidental and designed ecosystem scale manipulation studies will be examined.

Incidental Ecosystem Scale Experiments

Incidental ecosystem scale experiments are the result of some sort of activity that influences ecological processes at the scale of the entire ecosystem. For bivalve-dominated systems, the most common example of this type of experiment is intensive cultivation and harvest of these animals and the ecosystem changes that occur as a result of this activity. Specifically (Table 25), mussel culture in New Zealand changes the sediments and the nitrogen budget of the system (Kaspar et al. 1985), mussel culture in the Rias of Spain changes the sediment fauna and plankton (Tenore et al. 1982), oyster culture changed the carbon budget for Killary Harbor (Rodhouse and Roden 1987), oyster culture and overharvesting resulted in the loss of a cultured bivalve species and carbon budget changes in Marennes-Oléron (Héral 1993), and overharvesting and nutrient runoff resulted in a switch from a benthic-dominated to a pelagic-dominated system in Chesapeake Bay (Newell 1988). The intentional and unintentional (invasions) introduction of bivalve species and their subsequent population explosions in the San Francisco Bay has resulted in major ecosystem level changes (Cloern 1982; Nichols 1985; Carlton et al. 1990; Nichols et al. 1990; Alpine and Cloern 1992; Kimmerer et al. 1994). The major engineering changes to the Oosterschelde and the adjacent delta works in the Netherlands has had only a small influence on the role of bivalve-dominated ecosystem processes (Nienhuis and Smaal 1994). All of these cases have been examined previously in Chapters 5 and 6. Because Chesapeake Bay has been the focus of a long-term change and is the subject of ongoing studies across the range of ecosystem scale, it will be revisited here.

As a fishery, oysters, *Crassostrea virginica*, reached peak production in Chesapeake Bay in 1884. As recently as 1992, the production had declined to about 2% of the maximum production (Rothschild et al. 1994). These authors attribute the decline in oysters to habitat destruction and overfishing, and others have also implicated the additional effects of declining water quality as a result of pollution and increasing incidents of disease (Newell 1988). Some of the major ecological effects of the oyster fishery's decline on the Bay were already obvious, i.e., uneaten and dead phytoplankton causing increased bacterial abundance and anoxic water conditions, in the 1930s (Rothschild et al. 1994). Newell (1988), as discussed earlier, has argued that the decline in the oysters has led to the increase in planktonic suspension-feeding organisms with their attendant pelagic food web. He further argued that increasing the oyster population in the Bay through management and aquaculture could significantly improve water quality and allow the system to revert to its benthic dominated state. Newell's ideas were explored by Ulanowicz and Tuttle (1992) through the use of a simulation model (Figure 47) and through a preliminary and rough ecosystem level field experiment. The experiment compared the areas in the

TABLE 25
Direct and Incidental Ecosystem Manipulations That Result In Changes to Bivalve Mediated Processes

Ecosystem	Bivalve	System Change	Process or Structure Changed	Ref.
Chesapeake Bay, U.S.A.	*Crassostrea virginica*	Overharvesting	Switch from benthic to pelagic food webs	Newell 1988
Kenepunu Sound, New Zealand	*Perna canaliculus*	Intense cultivation	Higher nitrogen concentrations in sediments, more intense nitrogen cycling	Kaspar et al. 1988
Killary Harbour, Ireland	*Mytilus edulis*	Intense cultivation	Switch from pelagic to benthic bivalve filter feeding system	Rodhouse and Roden 1987
Marennes-Oléron, France	*Crassostrea angulata* *Crassostrea gigas* *Mytilus edulis*	Over-cultivation	Low growth, epidemics, extinction	Héral 1993
Oosterschelde, The Netherlands	*Cerastoderma edule* *Mytilus edulis*	Storm surge barrier constructed	No change	Nienhuis and Smaal 1994
Ria De Arosa, Spain	*Mytilus edulis*	Intense cultivation	Organically enriched sediments, high sediment biodiversity	Tenore et al. 1982
San Francisco Bay, U.S.A.	Various bivalves	Drought and high salinity, species invasion	Intense grazing of phytoplankton and zooplankton	Cloern 1982 Alpine and Cloern 1992 Nichols 1985 Kimmerer et al. 1994
North Inlet, U.S.A.	*Crassostrea virginica*	Oysters removed experimentally	Nutrient concentrations reduced in water	Dame and Libes 1993

Bay with oyster aquaculture plots to adjacent areas without such plots. They found that phytoplankton standing stocks in culture plots were about half those in uncultivated areas. Even though this experiment was of short duration and limited spatial scope, Ulanowicz and Tuttle (1992) felt it supported the potential increased filtration effects to be seen if oyster populations in the Bay were rehabilitated. However, just adding or increasing oysters to the present system may not switch it back to its original state. Gerritsen et al. (1994) calculated the capacity of benthic bivalves to filter the phytoplankton in the Bay, using the dominant species as of 1988. These species were *Corbicula fluminea*, *Macoma* spp., *Mulinia lateralis*, *Mya arenaria*, *Rangia cuneata*, and *Tagelus plebeius*. Notice that oysters are not listed. Using a probabilistic model of bivalve suspension feeding, these investigators found that bivalves could consume more than 50% of annual primary production in the shallow freshwater and oligohaline reaches of upper Chesapeake Bay and the Potomac River. In the deeper meosohaline potions, the model estimated that these bivalves could only consume about 10% of the primary production. As a result of their calculations, Gerritsen et al. (1994) argued that just adding oysters to the Bay would only have a limited effect in improving water quality, as they are mainly influencing the shallow portions of the Bay. Their study raises another ecological issue that is not discussed by any of these authors. Have the currently dominant bivalve filter feeders in Chesapeake Bay always been there at their current densities or have they risen to prominence as the oysters declined? Is this a case of functional redundancy with the Bay? As with all experiments, planned and otherwise, new and interesting questions arise from them.

Designed Ecosystem Scale Experiments

Designed experiments on aquatic ecosystems can be traced to the forested watershed studies of Likens et al. (1970) and the whole-lake experiments of Schindler (1990) and Carpenter and Kitchel (1993). In each of these large studies, major trophic components were manipulated and the ensuing changes on the ecosystem were monitored. As previously discussed, the 2-month pilot study by Dame and Libes (1993) investigated nutrient processing by oyster reefs in North Inlet, SC by the *in situ* experimental manipulation of tidal creeks through the removal of oysters and is the only documented study of this kind involving bivalves. To summarize, this summer season study found that the presence of oysters in tidal creeks increased the concentrations of inorganic nutrients and did not change the chlorophyll concentrations. This study is now (1996 to 1998) being repeated in a much more comprehensive mode, and the experimental design and considerations will be discussed below.

The interpretation of ecosystem level experiments requires that two questions be answered: (1) did the manipulated ecosystem change following the manipulation, and (2) did the manipulation cause the change? (Frost et al. 1988). These questions can be answered by randomized, replicated designs (Hurlbert 1984; Frost et al. 1988) and with less statistical power by other techniques such as intervention analysis (Carpenter et al. 1989). The BACI (before-after-control-impact) experimental design of Stewart-Oaten et al. (1986) assumes that observations are made on a potential site for some time before intervention or manipulation, and also on another site with similar characteristics. It is not necessary that the two sites match perfectly with respect to the distributions of the ecological parameters measured. After the intervention is applied to the randomly chosen experimental site, the change in mean value for any given parameter there is compared to any concomitant (natural) change at the untreated control site. This difference, referred to as the effect of the manipulation, may be immediate or gradual; it may also be either lasting or temporary. Underwood (1994) expanded upon the BACI idea by introducing multiple control sites and treating these as random effects in a multifactor mixed model analysis of variance.

If there are significant time-treatment interactions, then a staircase approach should be considered (Walters et al. 1988). In this less statistically powerful method, the manipulation, i.e., removal of oysters from a creek, is assumed to result in different effects, depending on the season of the removal. Thus the design must stage the removals to the seasons, with each removal compared to a control or unmanipulated system.

As ecosystem level manipulation experiments are expensive to replicate, it is often the case (see "Incidental Scale Experiments" above) that only one manipulated system and one reference system are available. In this case, Carpenter et al. (1989) have developed RIA (random intervention analysis) as a method to determine if nonrandom change has occurred in a manipulated ecosystem.

Using these statistical guidelines, we are now in the midst of a series of manipulation experiments on North Inlet that include the removal of oysters from creeks and the monitoring of ecosystem level changes over a 1-year postmanipulation period. Ecosystem processes under study are top-down and bottom-up control of plankton, nutrient dynamics, bacterio-plankton, phytoplankton, zooplankton and nekton biodiversity, total system metabolism and productivity, and physical dynamics of water exchange. In addition, network analysis is being applied for the first time to an experimental ecosystem manipulation. Within several years, the North Inlet team should have a much clearer and more comprehensive understanding of the role of oysters in the tidal creeks of this estuary.

The number of ecosystem scale field experiments involving marine bivalves is relatively few and this aspect is probably due to the expense involved in comprehensive monitoring and experimental replication. The incidental and designed studies have supported with simulations and mesocosms the lower scale investigations of nutrient cycling and grazing, but have also pointed out important areas of exploration that are not easily simulated, i.e., species invasions and biodiversity. The future begs for more comprehensive ecosystem scale studies that integrate computer simulations, mesocosms, and field manipulations to provide a clearer understanding of the ecosystem role of bivalves.

REFERENCES

Alpine, S.E. and Cloern, J.E. 1992. Trophic interactions and direct physical effects control phytoplankton biomass and production in an estuary, *Limnol. Oceanogr.*, 37, 946–955.

Bacher, C. 1989. Capacite trophique du bassin de Marennes-Oléron: couplage d'un modele de transport particulaire et d'un modele de croissance de l'huitre *Crassostrea gigas*, *Aquat. Living Resour.*, 2, 199–214.

Bacher, C., Héral, M., Deslous-Paoli, J.M., and Razet, D. 1991. Modele energetique uniboite de la croissance des huitres (*Crassostrea gigas*) dans le bassin de Marennes-Oléron, *Can. J. Fish. Aquat. Sci.*, 48, 391–404.

Baird, D., McGlade, J.M., and Ulanowicz, R.E. 1991. The comparative ecology of six marine ecosystems, *Phil. Trans. R. Soc. Lond. Ser. B*, 333, 15–29.

Baird, D. and Ulanowicz, R.E. 1989. A seasonal dynamics of the Chesapeake Bay ecosystem, *Ecol. Monogr.*, 59, 329–364.

Boucher, G. and Boucher-Rodoni, R. 1988. In situ measurement of respiratory metabolism and nitrogen fluxes at the interface of oyster beds, *Mar. Ecol. Prog. Ser.*, 44, 229–238.

Carlton, J.T., Thompson, J.K., Schemel, L.E., and Nichols, F.H. 1990. Remarkable invasion of San Francisco Bay, California, USA by the Asian clam *Potamocorbula amurensis*. I. Introduction and dispersal, *Mar. Ecol. Prog. Ser.*, 66, 81–94.

Carpenter, S.R., Chisholm, S.W., Krebs, C.J., Schindler, D.W. and Wright, R.F. 1995. Ecosystem experiments, *Science*, 269, 324–327.

Carpenter, S.R., Frost, T.M., Heisey, D., and Kratz, T.K. 1989. Randomized intervention analysis and the interpretation of whole-ecosystem experiments, *Ecology*, 70, 1142–1152.

Carpenter, S.R. and Kitchell, J.F. 1993. *The Trophic Cascade in Lakes*, Cambridge University Press, New York, 385 pp.

Cloern, J.E. 1982. Does the benthos control phytoplankton biomass in South San Francisco Bay?, *Mar. Ecol. Prog. Ser.*, 9, 191–202.

Connell, J.H. 1972. Community interactions on marine rocky intertidal shores, *Annu. Rev. Ecol. Sys.*, 3, 169–192.

Connell, J.H. 1975. Some mechanisms producing structure in natural communities: a model and evidence from field experiments, in Cody, M.L. and Diamond, J.M., Eds., *Ecology and Evolution of Communities*, Belknap/Harvard University Press, Cambridge, MA, 460–490.

Dame, R.F. and Libes, S.L. 1993. Oyster reefs and nutrient retention in tidal creeks, *J. Exp. Mar. Biol. Ecol.*, 171, 251–258.

Dame, R.F. and Patten, B.C. 1981. Analysis of energy flow in an intertidal oyster reef, *Mar. Ecol. Prog. Ser.*, 5, 115–124.

Dame, R.F., Zingmark, R.G., and Haskin, E. 1984. Oyster reefs as processors of estuarine materials, *J. Exp. Mar. Biol. Ecol.*, 83, 239–247.

Day, J.W., Smith W.G., Wagner, P.R., and Stone, W.C. 1973. *Community Structure and Carbon Budget of a Salt Marsh and Shallow Bay Estuarine System in Louisiana*, Center for Wetlands Resources, Louisiana State University, Baton Rouge.

DeLaune, R.D., Patrick, W.H., and Brannon, J. 1976. *Nutrient Transformations in Louisiana Salt Marsh Soils*, Center for Wetland Resources, Louisiana State University, Baton Rouge.

DiToro, D.M., O'Conner, D.J., and Thomann, R.V. 1971. A dynamic model of the phytoplankton in the Sacramento-San Joaquin delta, *Adv. Chem. Ser.*, 106, 131–180.

Doering, P.H., Kelly, J.R., Oviatt, C.A., and Sowers, T. 1987. Effect of the hard clam *Mercenaria mercenaria* on benthic fluxes of inorganic nutrients and gases, *Mar. Biol.*, 94, 377–383.

Doering, P.H. and Oviatt, C.A. 1986. Application of filtration rate models to field populations of bivalves: an assessment using experimental mesocosms, *Mar. Ecol. Prog. Ser.*, 31, 265–275.

Doering, P.H., Oviatt, C.A., Beatty, L.L., Banzon, V.F., Rice, R., Kelly, S.P., Sullivan, B.K., and Frithsen, J.B. 1989. Structure and function in a model coastal ecosystem: silicon, the benthos and eutrophication, *Mar. Ecol. Prog. Ser.*, 52, 287–299.

Doering, P.H., Oviatt, C.A., and Kelly, J.R. 1986. The effects of the filter-feeding clam *Mercenaria mercenaria* on carbon cycling in experimental marine mesocosms, *J. Mar. Res.*, 44, 839–861.

Eilers, P.H.C. and Peeters, J.C.H. 1988. A model for the relationship between light intensity and the rate of photosynthesis in phytoplankton, *Ecol. Model.*, 42, 199–215.

Eilers, P.H.C. and Peeters, J.C.H. 1993. Dynamic behaviour of a model for photosynthesis and photoinhibition, *Ecol. Model.*, 69, 113–133.

Emerson, C.W., Grant, J., Mallet, A., and Carver, C. 1994. Growth and survival of sea scallops *Placopecten magellanicus*: effects of culture depth, *Mar. Ecol. Prog. Ser.*, 108, 119–132.

Finn, J.T. 1976. measures of ecosystem structure and function derived from the analysis of flows, *J. Theor. Biol.*, 56, 363–380.

Finn, J.T. and Leschine, T.M. 1980. Does salt marsh fertilization enhance shellfish production? An application of flow analysis, *Env. Manag.*, 4, 193–203.

Frost, T.M., DeAngelis, D.L., Bartell, S.M., Hall, D.J., and Hurlbert, S.H. 1988. Scale in the design and interpretation of aquatic community research, in Carpenter, S.R., Ed., *Complex Interaction in Lake Communities*, Springer-Verlag, New York, 229–260.

Gerritsen, J., Holland, A.F., and Irvine, D.E. 1994. Suspension-feeding bivalves and the fate of primary production: an estuarine model applied to Chesapeake Bay, *Estuaries*, 17, 403–416.

Graneli, L.E., Olsson, P., Carlsson, P., Graneli, W., and Nylander, C. 1993. Weak 'top-down' control of dinoflagellate growth in the coastal Skagerrak, *J. Plank. Res.*, 15, 213–237.

Grant, J., Dowd, M., Thompson, K., Emerson, C., and Hatcher, A. 1993. Perspectives on field studies and related biological models of bivalve growth and carrying capacity, in Dame, R.F., Ed., *Bivalve Filter Feeders and Marine Ecosystem Processes*, Springer-Verlag, Heidelberg, 371–420.

Grant, J., Hatcher, A., Scott, D.B., Pockington, P., Schafer, C.T., and Winters, G.V. 1995. A multidisciplinary approach to evaluating impacts of shellfish aquaculture on benthic communities, *Estuaries*, 18, 124–144.

Grassle, J.F., Elmgren, R., and Grassle, J.P. 1981. Response of benthic communities in MERL experimental ecosystems to low level chronic additions of #2 fuel oil, *Mar. Environ. Res.*, 4, 279–297.

Hairston, N.G., Sr. 1989. *Ecological Experiments: Purpose, Design, and Execution*, Cambridge University Press, New York, 370 pp.

Halfon, E. 1979. *Theoretical Systems Ecology*, Academic Press, New York, 516 pp.

Hall, C. and Day, J. 1977. *Ecosystem Modeling in Theory and Practice*, Wiley, New York, pp.

Hannon, B. 1973. The structure of ecosystems, *J. Theor. Biol.*, 41, 535–546.

Hannon, B.and Ruth, M. 1994. *Dynamic Modeling*, Springer-Verlag, 248 pp.

Hatcher, A., Grant, J., and Schofield, B. 1994. Effects of suspended mussel culture (*Mytilus* spp.) on sedimentation, benthic respiration and sediment nutrient dynamics in a coastal bay, *Mar. Ecol. Prog. Ser.*, 115, 219–235.

Héral, M. 1993. Why carrying capacity models are useful tools for management of bivalve molluscs culture, in Dame, R.F., Ed., *Bivalve Filter Feeders in Estuarine and Coastal Ecosystem Processes*, Springer-Verlag, Heidelberg, 453–477.

Héral, M. Deslous-Paoli, J.M., and Prou, J. 1986. Dynamique des productions et des biomasses des huitres creuses cultivees (*Crassostrea angulata* et *Crassostrea gigas*) dans le bassin de Marennes Oléron depuis un siecle, *ICES, Mariculture Committee*, F: 41 pp.

Herman, P.M.J. 1994. The analysis of a dynamic simulation model for the Oosterschelde ecosystem — a summary, in Nienhuis, P.H. and Smaal, A.C., Eds., *The Oosterschelde Estuary (The Netherlands): A Case Study of a Changing Ecosystem*, Kluwer, Dordrecht, 431–436.

Herman, P.M.J. and Scholten, H. 1990. Can suspension-feeders stabilise estuarine ecosystems?, in Barnes, M. and Gibson, R.N., Eds., *Trophic Relationships in the Marine Environment*, Aberdeen University Press, Aberdeen,. 104–116.

Hofmann, E.E., Klinck, J.M., Powell, E.N., Boyles, S., and Ellis, M. 1994. Modelling oyster populations. II. Adult size and reproductive effort, *J. Shellfish Res.*, 13, 165–182.

Hofmann, E.E., Powell, E.N., Klinck, J.M., and Saunders, G. 1995. Modelling diseased oyster populations. I. Modelling *Perkinsus marinus* infections in oysters, *J. Shellfish Res.*, 14, 121–151.

Hofmann, E.E., Powell, E.N., Klinck, J.M., and Wilson, E.A. 1992. Modelling oyster populations. III. Critical feeding periods, growth and reproduction, *J. Shellfish Res.*, 11, 399–416.

Horsted, S.J., Nielsen, T.G., Riemann, B., Pock-Steen, J., and Bjørsen, P.K. 1988. Regulation of zooplankton by suspension-feeding bivalves and fish in estuarine enclosures, *Mar. Ecol. Prog. Ser.*, 48, 217–234.

Hurlbert, S.J. 1984. Pseudoreplication and the design of ecological field experiments, *Ecol. Monogr.*, 54, 187–211.

Kaspar, H.F., Gillespie, P.A., Boyer, I.C., and MacKenzie, A.L. 1985. Effects of mussel aquaculture on the nitrogen cycle and benthic communities in Kenepuru Sound, Marlborough Sounds, New Zealand, *Mar. Biol.*, 85, 127–136.

Kimmerer, W.J., Gartside, E., and Orsi, J.J. 1994. Predation by an introduced clam as the likely cause of substantial declines in zooplankton of San Francisco Bay, *Mar. Ecol. Prog. Ser.*, 113, 81–93.

Klepper, O. 1989. A Model of Carbon Flows in Relation to Macrobenthic Food supply in the Oosterschelde Estuary (S.W. Netherlands), Ph.D. thesis, Wageningen Agricultural University, Wageningen, the Netherlands, 270 pp.

Klepper, O., van der Tol, M.W.M., Scholten, H., and Herman, P.M.J. 1994. SMOES: a simulation model for the Oosterschelde ecosystem. I. Description and uncertainty analysis, in Nienhuis, P.H. and Smaal, A.C., Eds., *The Oosterschelde Estuary (The Netherlands): A Case Study of a Changing Ecosystem*, Kluwer, Dordrecht, 437–451.

Kremer, J.N. and Nixon, S.W. 1978. *A Coastal Marine Ecosystem —Simulation and Analysis*, Springer-Verlag, New York. 217 pp.

Lalli, C.M. 1990. Introduction, in Lalli, C.M., Ed., *Enclosed Experimental Marine Ecosystems: A Review and Recommendations*, Coastal and Estuarine Studies 37, Springer-Verlag, New York, 1–6.

Leschine, T.M. 1979. Salt marsh nitrogen flow analysis: fertilization and allocation of biological productivity, *Woods Hole Oceanographic Institution, Tech. Rep., WHOI*-79-29.

Likens, G.E., Borman, F.H., Johnson, N.M., Fisher, D.W., and Pierce, R.S. 1970. Effects of cutting and herbicide treatment on nutrient budgets in Hubbard Brook watershed-ecosystem, *Ecol. Monogr.*, 40, 23–47.

Menge, B.A. 1976. Organization of the New England rocky intertidal community: role of predation, competition, and environmental heterogeneity, *Ecol. Monogr.*, 46, 355–393.

Newell, R.I.E. 1988. Ecological changes in Chesapeake Bay: Are they the result of overharvesting the American oyster, *Crassostrea virginica*?, in Lynch, M.P. and Krome, E.C., Eds., *Understanding the Estuary: Advances in Chesapeake Bay Research*, Chesapeake Research Consortium, Solomon's, MD, 536–546.

Nichols, F.H. 1985. Increased benthic grazing: an alternative explanation for low phytoplankton biomass in northern San Francisco Bay during the 1976–1977 drought, *Estuarine Coastal Shelf Sci.*, 21, 379–388.

Nichols, F.H., Thompson, J.K. and Schemel, L.E. 1990. Remarkable invasion of San Francisco Bay (California, USA) by the Asian clam *Potamocorbula amurensis*. II. Displacement of a former community, *Mar. Ecol. Prog. Ser.*, 66, 95–101.

Nienhuis, P.H. and Smaal, S.C. 1994. *The Oosterschelde Estuary: A Case Study of A Changing Ecosystem* (reprinted from Hydrobiologia 282/283), Kluwer, Dordrecht, 597 pp.

Nixon, S.W., Alonso, D., Pilson, M.E., and Buckley, B.A. 1980. Turbulent mixing in aquatic microcosms, in Giesy, J.P., Ed., *Microcosms in Ecological Research*, DOE Symp. Ser. 52, CONF 781101, NTIS, 818–849.

Odum, E.P. 1983. *Basic Ecology*, Saunders, Philadelphia, 613 pp.

Odum, H.T. 1983. *Systems Ecology*, Wiley, New York. 644 pp.

Olsson, P., Graneli, E., Carlsson, P., and Abreu, P. 1992. Structuring of a post spring phytoplankton community by manipulation of trophic interactions, *J. Exp. Mar. Biol. Ecol.*, 158, 249–266.

Oviatt, C.A., Buckley, B., and Nixon, S.W. 1981. Annual phytoplankton metabolism in Narragansett Bay calculated from survey field measurements and microcosm observations, *Estuaries*, 4, 167–175.

Pain, R.T. 1977. Controlled manipulations in the marine intertidal zone and their contributions to ecological theory, in The Changing Scenes in Natural Sciences, 1776–1976. *Acad. Nat. Sci. Philadelphia, Spec. Pub.*, 12, 245–270.

Patten, B.C. 1971. *Systems Analysis and Simulation in Ecology*, Vol. 1, Academic Press, New York, 607 pp.

Patten, B.C., Bosserman, R.W., Finn, J.T., and Cale, W.G. 1976. Propagation of cause in ecosystems, in Patten, B.C., Ed., *System Analysis and Simulation in Ecology*, Vol. 4, Academic Press, New York, 457–579.

Pilson, M.E.Q. 1990. Application of mesocosms for solving problems in pollution research, in Lalli, C.M., Ed., *Enclosed Experimental Marine Ecosystems: A Review and Recommendations*, Coastal and Estuarine Studies 37, Springer-Verlag, New York, 155–168.

Pilson, M.E.Q., Oviatt, C.A., and Nixon, S.W. 1980. Annual nutrient cycles in a marine microcosm, in Giesy, J.P., Ed., *Microcosms in Ecological Research*, DOE Symp. Ser. 52, CONF 781101, NTIS, Springfield, VA, 753–778.

Platt, T., Mann, K.H., and Ulanowicz, R.E. 1981. *Mathematical Models in Biological Oceanography*, UNESCO Press, Paris, 156 pp.

Powell, E.N., Hofmann, E.E., Klinck, J.M., and Ray, S.M. 1992. Modeling oyster populations. I. A commentary on filtration rate. Is faster always better?, *J. Shellfish Res.*, 11, 387–398.

Powell, E.N., Klinck, J.M., Hofmann, E.E., and Ray, S.M. 1994. Modeling oyster populations. IV. Rates of mortality, population crashes, and management, *Fish. Bull.*, 92, 347–373.

Prins, T.C., Escaravage, V., Smaal, A.C., and Peeters, J.C.H. 1995. Nutrient cycling and phytoplankton dynamics in relation to mussel grazing in a mesocosm experiment, *Ophelia*, 41, 289–315.

Raillard, O., Deslous-Paoli, J.M., Héral, M., and Razet, D. 1993. Modelisation du comportement nutritionnel et de la croissance de l'huitre japonaise *Crassostrea gigas*, *Oceanol. Acta*, 16, 73–82.

Riemann, B., Nielsen, T.G., Horsted, S.J., Bjørsen, P.K., and Pock- Steen, J. 1988. Regulation of phytoplankton biomass in estuarine enclosures, *Mar. Ecol. Prog. Ser.*, 48, 205–215.

Riemann, B., Sørensen, M., Bjørnsen, P., Horsted, S., Jensen, L., Nielsen, T., and Søndergaard, M. 1990. Carbon budgets of the microbial food web in estuarine enclosures, *Mar. Ecol. Prog. Ser.*, 65, 159–170.

Rodhouse, P.G. and Roden, C.M. 1987. Carbon budget for a coastal inlet in relation to intensive cultivation of suspension-feeding bivalve molluscs, *Mar. Ecol. Prog. Ser.*, 36, 225–236.

Rothschild, B.J., Ault, J.S., Goulletquer, P., and Héral, M. 1994. Decline of the Chesapeake Bay oyster populations: a century of habitat destruction and overfishing, *Mar. Ecol. Prog. Ser.*, 111, 29–39.

Ruardij, P. and Baretta, J.W. 1989. The construction of the transport submodel, in Baretta, J. and Ruardij, P., Eds., *Tidal Flat Estuaries*, Springer-Verlag, Berlin, 65–76.

Schlesinger, S.R.E., Crosbie, R.E., Gagne, R.E., Innis, G.S., Lalwani, C.S., Lich, J., Sylvester, R.J., Wright, R.D. Kheir, N., and Bartos, D. 1979. Terminology for model credibility, *Simulation*, 32, 103–104.

Schindler, D.W. 1990. Experimental perturbations of whole lakes as tests of hypotheses concerning ecosystem structure and function, *Oikos*, 56, 25–41.

Scholten, H. and van der Tol, M.W.M. 1994. SMOES: a simulation model for the Oosterschelde ecosystem. II. Calibration and validation, in Nienhuis, P.H. and Smaal, A.C., Eds., *The Oosterschelde Estuary (The Netherlands): A Case Study of a Changing Ecosystem*, Kluwer, Dordrecht, 453–474.

Smith, W., Gibson, V.R., and Grassle, J.F. 1982. Replication in controlled marine systems: representing evidence, in Grice, G.D. and Reeve, M.R., Eds., *Marine Mesocosms*, Springer-Verlag, New York, 217–225.

Stewart-Oaten, A., Murdoch, W.M., and Parker, K.R. 1986. Environmental impact assessment: "pseudoreplication" in time?, *Ecology*, 67, 929–940.

Tenore, K.R., Boyer, L.F., Cal, R.M., Corral, J, Garcia- Fernandez, C., Gonzalez, N., Gonzalez-Gurrianan, E., Hanson, R.B., Iglesias, J., Krom, M., Lopez-Jamar, E., McClain, J., Pamatmat, M.M., Perez, A., Rhoads, D.C., de Santiago, G., Tiejen, J., Westrich, J., and Windom, H.L. 1982. Coastal upwelling in the Rias Bajas, NW Spain: contrasting the benthic regimes of the Rias de Arosa and de Muros, *J. Mar. Res.*, 40, 701–772.

Ulanowicz, R.E. 1980. A hypothesis on the development of natural communities, *J. Theor. Biol.*, 85, 233–245.

Ulanowicz, R.E. 1986. *Growth and Development: Ecosystems Phenomenology*, Springer-Verlag, New York. 203 pp.

Ulanowicz, R.E. and Platt, T. 1985. Ecosystem theory for biological oceanography, *Can. Bull. Fish. Aquat. Sci.*, 213, 1–260.

Ulanowicz, R.E. and Tuttle, J.H. 1992. The trophic consequences of oyster stock rehabilitation in Chesapeake Bay, *Estuaries*, 15, 298–306.

Underwood, A.J. 1994. On beyond BACI: sampling designs that might reliably detect environmental disturbances, *Ecol. Appl.*, 4, 3–15.

Valiela, I., Teal, J.M., and Sass, W.J. 1975. Production and dynamics of salt marsh vegetation and the effects of experimental treatment with sewage sludge, *J. Appl. Ecol.*, 12, 973–982.

Vargo, G.A., Hutchins, M., and Almquist, G. 1982. The effects of low, chronic levels of no. 2 fuel oil on natural phytoplankton assemblages in microcosms. I. Species composition and seasonal succession, *Mar. Environ. Res.*, 6, 245–264.

Walters, C.J., Collie, J., and Webb, T. 1988. Experimental designs for estimating transient responses to management disturbances, *Can. J. Fish. Aquat. Sci.*, 45, 530–538.

de Wilde, P.A.W.J. 1990. Benthic mesocosms: I. Basic research in soft-bottom benthic mesocosms, in Lalli, C.M., Ed., *Enclosed Experimental Marine Ecosystems: A Review and Recommendations*, Coastal and Estuarine Studies 37, Springer-Verlag, New York, 109–121.

Wulff, F. and Ulanowicz, R.E. 1989. A comparative anatomy of the Baltic Sea and Chesapeake Bay ecosystems, in Field, J.G., Wulff, F., and Mann, K.H. 1989. *Network Analysis in Marine Ecology*, Springer-Verlag, New York, 232–256.

Zeitschel, B. and Davies, J.M. 1978. Benthic growth chambers, *Rapp. P.V. Reun. Cons. Int. Explor. Mer.*, 173, 31–42.

8 Bivalves as Components of Ecosystem Health

INTRODUCTION

Ecosystem health may be defined in terms of characteristics applicable to any complex system (Costanza et al. 1992): sustainability that is a function of activity, organization, and resilience. Thus, *an ecosystem is healthy and free from distress if it is stable and sustainable — that is, if it is active and maintains its organization and autonomy over time and is resilient to stress.* In this definition, it is implied that the ecosystem is sustainable and can maintain its structure and function over time.

As each ecosystem is unique, defining ecosystem health is a dynamic process that requires: (1) the identification of important indicators of health, i.e., metabolism, nutrient flux, primary production, etc.; (2) the categorizing of important endpoints of health, i.e., composites of indicators, species, and systems; and (3) the identification of a healthy state incorporating our values and overall system performance (Costanza 1992).

The concept of ecosystem health applies the model of human or animal medicine to the practice of "ecological medicine." The assessment of health in medicine is:

1. Identify the symptoms
2. Identify and measure the vital signs
3. Make a provisional diagnosis
4. Conduct tests to verify the diagnosis
5. Make a prognosis
6. Prescribe a treatment

While the model is applicable to ecosystems, ecologists do not have a compendium of known diseases or stresses with associated symptoms and signs. They have, however, begun to categorize system level stresses and record their effects (symptoms) on the ecosystem (Costanza et al. 1992). Table 26 provides a first attempt at presenting these characteristics for bivalve dominated systems. As more observations are added to this compendium, the next step is to develop quick diagnostic tests that will enable the ecologists to detect stress early (before ecosystem retrogression) and recommend remedial action. At this point, preventive ecosystem scale medicine can be practiced. In Chesapeake Bay, for example, current indications (see Chapters 5, 6, and 7) are that the bay is becoming hyper-eutrophicated. The symptoms of hyper-eutrophication are shorter food chains with a lower degree of organization, lower diversity, and higher productivity. These symptoms are further complicated by the

TABLE 26
Characteristic Responses (Symptoms) of Bivalve-Dominated Ecosystems to Stress

Stress	Nutrients	Primary Production	Size Distribution	Species Diversity	System Retrogression
Harvesting of renewable resources	–	–	–	–	+
Nutrient loading	+	+	–	–	+
Physical restructuring	*	*	–	–	+
Introduction of species	+	+	*	–	+
Extreme natural events	*	*	–	–	+

Note: Signs (+ or –) indicate the direction of change compared with normal functioning of relatively unstressed systems. (*) Response not observed as yet

Modified from Rapport et al. 1985.

loss of the major grazer in the system, oysters, due to overharvesting. Today, there is a multiscale management plan for Chesapeake Bay that addresses both of these major impacts on the bay and includes efforts to rehabilitate the oyster population (Costanza et al. 1992).

If the data base of symptoms and appropriate diagnostic tools for assessing ecosystem health are lacking, Costanza et al. (1992) recommend network analysis and simulation (Chapter 7) modeling as stopgap approaches. Network measures like ascendancy estimate how many different species there are in an ecosystem and how these species are organized. Simulation modeling allows the prediction of ecosystem responses to various management alternatives and to natural environmental changes.

In this chapter, the use of bivalves as indicator species for measures of health in estuarine and coastal marine ecosystems will be reviewed. The various approaches will be organized according to relative time scale, starting with the quicker responding techniques.

BIVALVES AS MONITORS

For almost a century, marine environmental monitoring focused on abiotic parameters like temperature, salinity, currents, and nutrients. With the advent of an awareness that pollutants (implies that something causes a biological effect) and contaminants (result from the discharge to the environment of materials in excess of normal concentrations) could affect mankind and the environment (Widdows and Donkin 1992), monitoring efforts were refocused on the detection of anthropogenic materials in order to obtain information on distribution, transport mechanisms, and fates of these materials. Initially, analytical techniques had to be perfected, but eventually these problems were overcome and routine abiotic monitoring became widespread. However, the discrete sampling of natural waters followed by chemical analysis is considered inadequate for three reasons (Baldwin and Kramer 1994):

1. It fails to identify contaminants which are not included in the routine analysis but which may nonetheless occur.
2. It does not necessarily relate directly to the impact that measured chemicals have on the biota and ecosystem.
3. It may fail to identify sporadic contamination of the ecosystem.

In order to gain more information, continuous monitoring and biological monitoring have been developed. A number of attributes of bivalves have led to their use as "monitors," "sentinels," or "indicators" of environmental stress (Widdows and Donkin 1992; Smaal and Widdows 1994):

1. Bivalves are dominant members of coastal and estuarine systems and have wide geographical distributions.
2. These animals are sedentary and serve as integrators of contamination in a specific area.
3. They are relatively tolerant of (but not insensitive to) a wide range of environmental conditions.
4. Suspension feeding bivalves pump large volumes of water and concentrate many chemicals by several orders of magnitude over concentrations in seawater.
5. Concentrations of contaminants in bivalve tissues provide an assessment of biological availability.
6. Bivalves have a very low level of activity for those enzyme systems capable of metabolizing organic contaminants.
7. Commercially important bivalve populations are relatively stable.
8. Bivalves can be readily transplanted and maintained in specific sites.
9. Several bivalves are commercially important as seafood, and their chemical contamination is of interest to public health.
10. Bivalves have several behavioral and physiological responses to stress that are easily and quickly measured.

With so many advantages, it is not surprising that bivalves have been successfully used as short-term and long-term monitors (Table 27) of environmental stress in coastal and estuarine waters.

BIVALVES AS INTEGRATORS

The direct assessment of chemical pollution is difficult, because pollutant concentrations in seawater are often so low that chemical pre-concentration of samples is necessary before analysis. In other words, the concentration of pollutants is often below detection limits. As a result, clean and exacting water sample collection and handling techniques must be employed. In addition, the inherent variability of pollutant concentrations in seawater due to temporal and spatial variability in sources and/or removal processes require frequent sampling to assess mean or time-averaged levels (Pitts and Wallace 1994). As an alternative, chemical concentrations in various bivalve tissues (soft and hard) are often used as indicators of ambient water chem-

TABLE 27
Temporal Scales of Various Bivalve Monitoring Approaches for Assessing Ecosystem Health

Organizational Level	Process	Measurement	Stress Response	Temporal Scale
Organism	Behavioral	Shell movement	Gaping orextended closure	sec–min
Organism	Physiological	Growth	Reduced	weeks–years
Organism	Physiological	Condition index	Decrease	days–weeks
Organism	Physiological	Anaerobic metabolism	Increase	min–hours
Organism	Physiological	Pollutant or contaminant concentration	Increase in shell or tissue	weeks–years
Organism–population	Physiological	Reproduction	Decrease	weeks–years
Organism–population	Infection	Parasitism or disease	Increase	days
Organism–population	Toxicological	Mortality	Death	days
Organism–population	Metabolic	Scope for growth	Decrease	days–months
Community	Metabolic	Production: respiration	Decrease or increase	days–years
Ecosystem	Metabolic	Production: respiration	Decrease or increase	hours–years
Ecosystem	Connectivity	Ascendancy	Decrease	months–years

istry; because of their feeding processes, these organisms can concentrate many chemical species by orders of magnitude over seawater concentrations. Furthermore, pollutant levels in specific tissues provide an indication of bioavailability (Pitts and Wallace 1994). For these and previously listed reasons, bivalves have been used as monitors of chemical contamination in coastal waters, particularly in areas receiving waste discharged from point sources such municipal waste treatment plants or non-point sources such as storm water runoff.

There is strong experimental evidence supporting the use of living bivalves as biomonitors. Numerous laboratory experiments with controlled concentrations and exposures (Cunningham and Tripp 1975; Fisher and Teyssie 1986; etc.), field transplant experiments of oysters and mussels between high and low concentration areas (Roesijadi et al. 1984; Martin 1985, etc.) have demonstrated the accumulation of pollutants in bivalves. The degree to which inorganic and organic contaminants are accumulated by bivalves depends on both abiotic physiochemical properties, and biotic factors, such as pumping activity, growth, biochemical composition, reproductive condition, and metabolism. All of these various factors influence the rates of the dynamic processes concerned with uptake, deposition, and depuration, that together determine the degree of bioaccumulation (Widdows and Donkin 1992).

As early as 1976, the U.S. initiated the "Mussel Watch Monitoring Program" as a means to monitor pollution in coastal waters using mussels and oysters (Goldberg et al. 1978). Similar programs were developed in the U.K., France, Canada, Australia, Japan, Taiwan, India, South Africa, and the soviet republics. In these early efforts, bivalve body tissues were analyzed for four classes of pollutants: trace metals, chlorinated hydrocarbons, petroleum hydrocarbons and radionuclides (Goldberg et al. 1978, 1983). The primary motivation of these observations was to protect human health by estimating exposure via the diet back to humans and protect valuable living resources (Widdows and Donkin 1992).

The early version of Mussel Watch in the U.S. was eventually superseded by the National Status and Trends (NS&T) Mussel Watch Project in 1986. This project annually samples more than 100 sites along the Atlantic, Gulf, and Pacific coasts of the U.S. *Mytilus edulis* is the sampled species from Maine to Delaware Bay, *Crassostrea virginica* from Delaware Bay through the Gulf of Mexico, and the mussels *Mytilus edulis* and *Mytilus californianus* along the West Coast. The chemicals analyzed for are DDT and its metabolites (6), chlorinated pesticides other than DDT (9), polychlorinated biphenyls (18 congeners), polycyclic aromatic hydrocarbons (24), the major elements aluminum, iron, and manganese, and trace elements (11). Except for the major elements, all are possible contaminants in the sense that their concentrations in the environment have been altered by human activities (O'Connor et al. 1994).

In monitoring programs over wide geographic ranges, several bivalve species may be used, but species intercomparability can be a problem. For example, oysters and mussels are not equal in their ability to concentrate chemicals. When the two bivalves are compared at the same site, the trace elements of silver, copper, and zinc are highly enriched by oysters, while chromium and lead are higher in mussels (O'Connor et al. 1994). No strong species effect was found for organic compounds or other elements. Similar species, i.e., *M. edulis* and *M. californianus*, concentrated chemicals similarly. In order to offset physiological effects within a bivalve species, samples are taken on a set size range, at the same site, and in the winter prior to the spawning season.

The on-going NS&T Mussel Watch Project analyzes trends through the application of nonparametric statistics to raw concentration data, with no attempt to adjust these concentrations for physiological or external factors. For the U.S., few trends have been detected, and those that have been identified are predominantly decreases. As trend detection will improve with time (the number of observations increase), it is expected that trends now obscured by uncontrollable natural factors will emerge (O'Connor et al. 1994).

Recent studies by Pitts and Wallace (1994) have shown that in addition to the concentration of trace elements in the soft body tissues of bivalves, analysis of such elements in the bivalve shell is also feasible. Using the clam *Mya arenaria*, these investigators found that there was a strong correlation between lead in the shell and that dissolved in seawater. Their method requires few samples ($n = 5 - 10$) and can be done on individual annual growth layers (see Chapter 3). Thus, the shell analysis technique offers not only current estimates of lead in a given coastal environment, but also estimates lead concentrations from the recent and ancient past. This analysis is certainly worth consideration as an additional monitoring technique in the Mussel Watch suite.

Generally, the mussel watch type programs have not only quantified the degree of contamination, but they have identified unexpected contamination hot spots (Widdows and Donkin 1992). Thus, bivalves as sentinel organisms have proved to be useful tools in identifying variations in chemical contamination between sites, and have contributed to an understanding of trends in coastal contamination (Widdows and Donkin 1992).

BIVALVE RESPONSES

Widdows and Donkin (1992) have argued that pollution implies a biological effect, while contamination is a physical–chemical phenomenon. Therefore, they contend that the assessment of pollution and environmental quality must ultimately be in terms of biological measurements, preferably in concert with appropriate measurements of chemical contaminants.

Appropriate biological observations for assessing and monitoring environmental quality should meet most of the following criteria (Widdows and Donkin 1992):

1. They should be sensitive to actual environmental levels of pollutants, and have a response from optimal to lethal conditions.
2. They should reflect a quantitative and predictable relationship with toxic contaminants.
3. They should have a relatively short response time so that pollution impacts can be detected in their incipient stages.
4. The technique should be applicable to both laboratory and field studies in order to associate laboratory-based concentration–response relationships to field measurements of spatial and temporal changes in environmental quality.
5. They should not only provide an integrated response to the total pollutant load, but should also provide insight into the underlying cause and the mechanism of toxicity.
6. The biological response should have ecological relevance and be shown to reflect deleterious effects on fundamental physiological parameters of the individual which translate to the level of the population and the system.

Some responses of bivalves that have been used to measure the effects of contaminants and the effects of environmental pollution are: shell movement, mortality, shell growth, tissue growth and scope for growth.

SHELL MOVEMENTS

In a time perspective, most biological approaches to monitoring focus on the past occurrence of pollutants over weeks or longer time spans. If an immediate warning of a pollution event is needed, then some sort of continually monitored biological response must be considered. These “biological early warning systems” (BEWS) provide a rapid warning of the occurrence of contaminants at concentrations that could be immediate threats to living systems. These systems are a combination of an automated continuous biological monitor serving as a sensing element and coupled to an electrical, optical, or mechanical sensing element that transfers the information to a monitoring computer. They usually measure a physiological or behavioral response of the organism (Baldwin and Kramer 1994).

A number of physiological responses by bivalves have been tested in the laboratory for their suitability to detect environmental change (Akberali and Trueman 1985). A few that seem promising for use in BEWS systems are heart beat, respiration, filtration, pumping rate, and shell or valve closure (Baldwin and Kramer 1994). Of these only shell closure has, to date, led to a practical BEWS technique. Closure of the valves is a typical example of an escape or avoidance behavior response, as bivalves normally have their shells open for feeding and respiration. Numerous studies have shown that bivalves close their shell under both natural and anthropogenic stress, and this closure may last for an extended period of time. Certain other stresses, particularly pollutants such as chlorine and organic solvents, cause a dramatic increase in valve movement activity (Kramer et al. 1989).

A BEWS has been developed using the valve movement response in blue mussels. This system is commercially known as the Musselmonitor (Kramer et al. 1989). The mussel monitor is designed to operate *in situ* and is contained in a waterproof housing. Valve movement is detected with a high frequency electromagnetic induction system. The electronic sensor consists of two small coils, glued to opposite shell halves of the mussel. One coil acts as the transmitter, generating a magnetic field, and the other coil functions as a receiver. Across the width of the individual bivalve, a linear response is obtained. The system is normally deployed with eight mussels, and each mussel is individually monitored to discriminate individual mussel behavior and improve the reliability of alarm detection. Valve movement likewise allows the detection of animal death and low battery power. This system also permits optimization for maximum sensitivity and minimum false alarm (Baldwin and Kramer 1994). As the system is operating *in situ*, handling is minimized, and the system has been deployed for more than 3 months without replacement of the mussels. Filter feeding bivalves are considered best for use in the mussel monitor because they are in maximum contact with the water column. *Mytilus edulis* and *Ostrea* spp. have been used in temperate estuarine situations, while the green mussel, *Perna viridis*, offers possibilities in tropical areas (Baldwin and Kramer 1994).

BEWS are thought to be best used when a relatively steep gradient in pollution concentration is to be detected. In marine conditions, dilution may be a factor, and in estuarine systems, salinity may interfere with the measurements. It is thought BEWS should be useful for effluent monitoring near industrial effluent lines, offshore discharges, aquaculture sites, and intake water for public aquaria and marinas (Baldwin and Kramer 1994).

Mortality

In the laboratory, the most common measure of acute toxicity to chemical contaminants is the 96 h LC_{50} test, which determines the toxicant concentration resulting in a 50% lethal response over a period of 96 h exposure. This approach works well with larval bivalves (populations), but leads to false estimates of tolerance in adult bivalves because they usually close their valves and thus isolate themselves from extreme environmental conditions for long periods of time. Also, it takes a long time for adult bivalve mortality effects to overcome extensive body reserves, but long-term lethal effects can be predicted from measurements of negative energy balance or scope for growth (Widdows and Donkin 1992). The mortality approach is seldom advocated because it lacks the sensitivity either to protect the environment by adequately screening toxic materials prior to release, or to identify pollutant effects in all but the most extreme and acute environmental pollution events.

A more promising approach is the resistance of bivalves to aerial exposure (Eertman and de Zwann 1994). Some bivalves, like mussels, when acclimated to subtidal conditions are unable to consume enough oxygen when suddenly exposed to air. Thus, aerial exposure of these bivalves, like exposure to environmental anoxia, will gradually result in their tissues becoming anoxic and in their metabolic processes switching from aerobic to anaerobic. In those species that are able to consume oxygen during aerial exposure, anoxia tolerance can be determined by exposing these bivalves to anoxic seawater.

Numerous studies (see Eertman and de Zwaan 1994) indicate that natural and/or anthropogenic factors that reduce anoxia tolerance will lead to higher mortality or reduced fitness in bivalves. Such factors are stress indicators according to Bayne's (1980) definition. Anoxic survival is, consequently, a stress index where the damage caused by environmental factors is expressed as a reduced capability to resist environmental change in the form of blocked supply of oxygen. This stress is evaluated from mortality data. The method is simple and inexpensive, with general applicability potential to coastal monitoring programs.

Shell and Tissue Growth

The rate of growth is a basic measure of the physiological fitness of an organism and, to a certain extent, a population. For this reason, growth has frequently been used as a measure of pollution effects and environmental quality. In bivalves, growth is often difficult to measure in a quantifiable way because: (1) production due to reproduction (gametes) is lost from the animal and extremely difficult to determine directly; (2) weight changes are difficult to measure because of shells and trapped water within shell; (3) there is no tight coupling between shell growth and that of soft tissues and gonads; and (4) it is difficult to separate nutrition and sublethal toxicant effects in field measurements of growth (Widdows and Donkin 1992). The end result of these problems is that natural environmental variability in growth can mask sublethal effects of pollutants.

The preceding concerns are less of a problem when bivalve larvae are the organisms selected, as they are followed in the laboratory under controlled conditions for short time periods and growth measurements are usually combined with mortality estimates (Calabrese et al. 1977; Widdows 1991).

Bivalve shell shape may also be indicative of certain pollutants. Certain antifouling paints release chemicals that cause shell thickening and shell deformities in oysters and mussels (Waldock and Thain 1983; Widdows and Donkin 1992).

Finally, both tissue and shell measures have been combined in condition indices (see Chapter 3) that relate tissue wet weight to shell volume or total wet weight. Although this approach has been used by shellfish biologists to determine the condition of animals in relation to seasonal and spatial changes (Widdows and Donkin 1992), it has had almost no success in estimating pollution effects.

Bivalve Scope For Growth (SFG)

The growth process in bivalves can be disrupted by changes in environmental quality and stress. Scope for growth (SFG), as described in Chapter 3, is an integrated physiological parameter that reflects the balance between the processes of energy acquisition (feeding and absorption) and energy expenditure (metabolism and excretion). This determination of the energy available for growth is estimated by an energy budget, rather than by the direct measurement of growth itself, and has proven to be particularly effective in assessing the biological effects of pollution (Widdows and Johnson 1988; Smaal and Widdows 1994). The SFG approach provides not only an instantaneous estimate of the energy status of the animal, but also supplies insight into the underlying physiological components that effect changes in the growth rate. In addition, there is supporting evidence that there is agreement between indirect growth estimates using energy budgets and direct methods measuring tissue and shell growth (Riisgård and Randløv 1981), and observations of production utilizing detailed population analysis (Gilfillan and Vandermenlen 1978; Bayne and Worrall 1980; Grant and Cranford 1991).

Scope for growth can be defined by the energy budget equation:

$$C = P + R + E + F$$

and

$$P = A - (R + E)$$

where C is consumption, P is total production (shell, tissues, gametes), R is respiration, E is excretion, F is fecal energy, and A is assimilated or absorbed energy. When production is estimated by the difference between assimilation and respiration plus excretion it is called scope for growth (Warren and Davis 1967). When scope for growth is estimated under well-defined conditions, it provides an integration of the various physiological parameters involved and a discrimination between natural and anthropogenic stress (Figure 50).

The advantages of using the scope for growth estimate on bivalves in marine ecosystems as an environmental quality monitoring technique are (Smaal and Widdows 1994):

1. SFG provides a sensitive, quantitative, and integrated stress response over a wide range of conditions.
2. A decline in growth has ecological relevance and is readily interpretable as a deleterious effect.
3. SFG is an early warning indicator of cellular and metabolic responses at the individual level, which allows detection of responses to environmental change prior to measurable effects on the growth, reproduction, or survival of the individual.
4. SFG represents an instantaneous estimate of growth potential which can be applied rapidly and cost effectively both in the laboratory and under field situations.

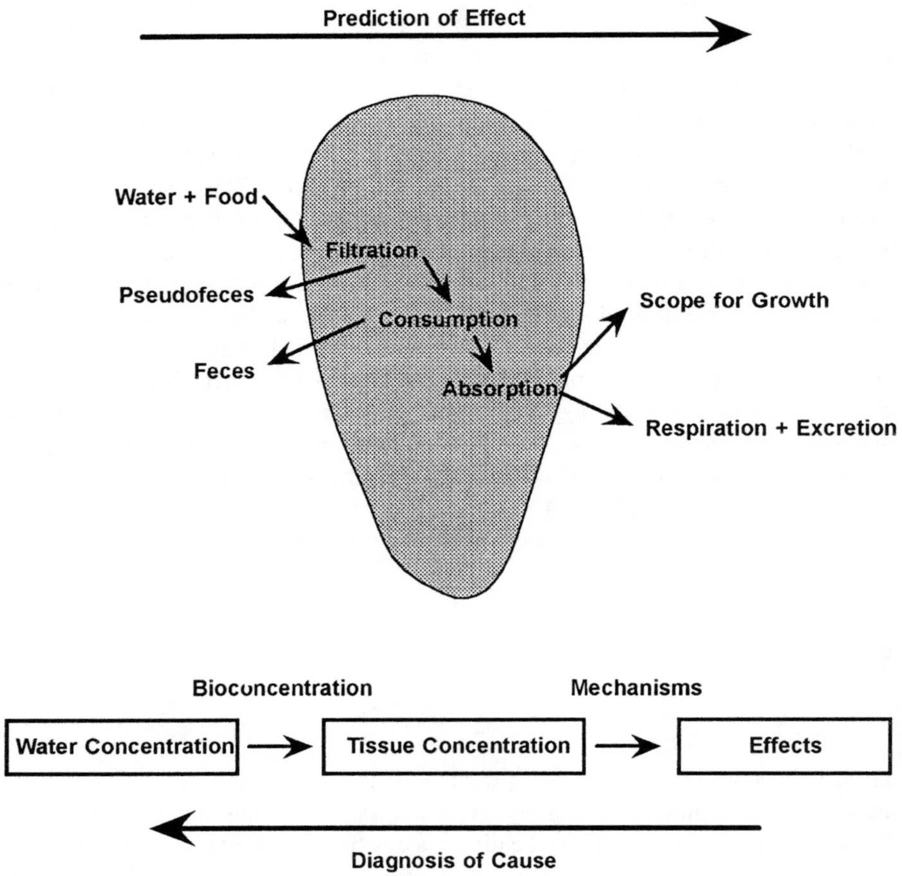

FIGURE 50. Bivalve scope for growth as part of the original energy budget with an ecotoxicology framework. (Adapted from Widdows and Donkin 1992.)

The disadvantages of the SFG approach include initial equipment costs and a need for more highly trained technical capabilities for the personnel involved. For a discussion of specific techniques for using SFG on bivalves, see Smaal and Widdows (1994).

A number of both natural and anthropogenic factors influence the estimation of scope for growth in bivalves. The physiological rates of feeding, respiration, and excretion all increase with increasing body size in an allometric manner (see Chapter 3). Reproductive condition can influence SFG as respiration and excretion rates are higher when gonads are fully developed (Smaal and Widdows 1994). Hence, the spawning period should be avoided in SFG measurements. Genetic differences have only a slight influence on SFG, but parasites can reduce filtration rates and growth rates. Laboratory conditions can affect the SFG of bivalves when these conditions exceed the adaptive capacity of field-acclimatized animals. For this reason, laboratory conditions should be close to ambient field conditions (Smaal and Widdows 1994).

SFG in bivalves can be influenced directly, as an acute response to chemical contaminants in water and sediment (Smaal and Widdows 1994), or after chronic exposure and long-term accumulation of pollutants (Widdows and Donkin 1992). Numerous field studies show that SFG varies significantly in mussels collected along estuarine or pollution gradients, i.e., there is lower SFG with higher pollution. Those interested in details of various contaminants and their influence on bivalve SFG may see Smaal and Widdows (1994).

The scope for growth approach for quantifying the biological effects of pollutants has several important attributes (Widdows and Johnson 1988).

1. It demonstrates sensitivity to environmentally realistic levels of pollution.
2. It reflects a quantitative and predictable relation with the concentrations of toxicants in the bivalve body tissues.
3. It represents a general response to the total pollution stimulus in a given system.
4. Bivalve feeding and growth responses are readily interpretable in terms of significant negative effects at the level of the organism and the population.

At present, it appears difficult to extrapolate SFG to the health of the ecosystem, but the theoretical foundation is developing (Costanza 1992).

SYSTEM MEASURES

As this book has tried to show, bivalves can be integral and important components in marine and estuarine ecosystems. While these animals are good monitors of pollution and stress in ecosystems, the transition from their organismic and population status as indicators of environmental quality to signs of ecosystem health is at best vague, because these indicators do not reflect the complexity of ecosystems. Costanza (1992) has argued that there is a progression from directly measured indicators of a component's status, through endpoints that are composites of these indicators, to measures of system values including performance and health. While indicators are directly measured tangible observations of generally high precision, values are difficult to determine, highly comprehensive, and relevant. As we have shown, there are a number of bivalve processes that are good indicators of environmental quality and in some systems bivalve populations and their associated communities are important enough to be considered endpoints. However, the final step to values or overall system health, the most integrative step, has yet to be accomplished.

To meet the goal of a practical definition of ecosystem health, Costanza (1992) believes that the concept of ecosystem health should include a combined measure of the system (Table 3): (1) resilience, ability to maintain its structure and patterns of behavior in the face of disturbance; (2) balance, proper equilibrium between system components; (3) organization, diversity and complexity; and (4) vigor, system metabolism or scope for growth — the difference between the energy required for system maintenance and the energy available to the system for all purposes (P/R).

Other considerations include a comprehensive description of the system, a weighting of factors to compare and aggregate the different components, and a hierarchical presentation to account for the interdependence of various time and space scales.

Thus, Costanza (1992) proposes a preliminary index of overall system health:

$$HI = V \cdot O \cdot R$$

Where HI = system health index, and also indicates sustainability; V = system vigor; O = system organization; and R = system resilience. In Table 28 the probable methods of determining the various components of HI are listed. It is interesting that all of these approaches are currently being pursued in ecosystems where bivalves are a major component and which have been reviewed earlier in this text. It would not be surprising, therefore, to see ecosystems dominated by bivalves at the forefront of quantifying and defining ecosystem health.

TABLE 28
Potential Components of an Ecosystem Health Index as Envisioned by Costanza

System Health Component	Related Concept	Related Measures	Method
Vigor	Function Productivity Throughput	GPP, NPP, P/R, metabolism	Measurement
Organization	Structure Biodiversity	Diversity index	Network analysis
Resilience	Disturbance	Scope for growth	Simulation
Combinations	Connectivity Complexity	Ascendancy	Simulation

From Costanza 1992.

CONCLUSIONS

From the preceding discussion, bivalves, particularly mussels, oysters, and clams, have a number of the characteristics of good indicator and monitoring species. Using the ecological medicine model described in the introduction to this chapter, bivalves can and currently do play an important role in the determination and maintenance of ecosystem health in many coastal ecosystems. Because these animals are proven integrative monitors, they are commonly used as indicators of numerous symptoms of system stress and may be important components in determining the vital signs of some ecosystems. From the early warning of system problems, bivalves can be instrumental in developing provisional diagnoses and providing reliable test subjects to verify these diagnoses. After a treatment program is initiated, bivalves can be important monitors of its effectiveness. That these bivalves are also often major components of their ecosystems also promises to ensure that they and their systems will play a prominent role in the development of ecosystem health indices and values, not only in coastal and estuarine systems, but in ecosystems in general.

REFERENCES

Akberali, H.B. and Treuman, E.R. 1985. Effects of environmental stress on marine bivalve molluscs, *Adv. Mar. Biol.*, 22, 102–199.

Baldwin, I.G. and Kramer, K.J.M. 1994. Biological early warning systems (BEWS), in Kramer, K.J.M., Ed., *Biomonitoring of Coastal Waters and Estuaries*, CRC Press, Boca Raton, FL, 1–28.

Bayne, B.L. 1980. Physiological measurements of stress, *Rapp. P.-V. Reun. Cons. Int. Explor. Mer*, 179, 56–61.

Bayne, B.L. and Worrall, C.M. 1980. Growth and production of mussels (*Mytilus edulis*) from two populations, *Mar. Ecol. Prog. Ser.*, 3, 317–328.

Calabrese, A., MacInnes, J.R., Nelson, D.A., and Miller, J.E. 1977. Survival and growth of bivalve larvae under heavy-metal stress, *Mar. Biol.*, 41, 179–184.

Costanza, R. 1992. Toward an operational definition of ecosystem health, in Costanza, R., Norton, B.G. and Haskell, B.D., Eds, *Ecosystem Health: New Goals for Environmental Management*, Island Press, Washington, D.C., 239–256.

Costanza, R., Norton, B.G., and Haskell, B.D., Eds. 1992. *Ecosystem Health: New Goals for Environmental Management*, Island Press, Washington, D.C., 269 pp.

Cunningham, P.A. and Tripp, M.R. 1975. Factors affecting the accumulation and removal of mercury from tissues of the American oyster *Crassostrea virginica*, *Mar. Biol.*, 31, 311–319.

Eertman, R.H.M. and de Zwaan, A. 1994. Survival of the fittest: resistance of mussels to aerial exposure, in Kramer, K.J.M., Ed., *Biomonitoring of Coastal Waters and Estuaries*, CRC Press, Boca Raton, FL, 269–284..

Fisher, N.S. and Teyssie, J.-L. 1986. Influence of food composition on biokinetics and tissue distribution of zinc and americium in mussels, *Mar. Ecol. Prog. Ser.*, 28:197–207.

Gilfillan, E.S. and Vandermenlen, J.H. 1978. Alterations in growth and physiology of soft shell clams, *Mya arenaria*: chronically oiled with Bunker C from Chedabucto Bay, Nova Scotia, 1970–1976, *J. Fish. Res. Board Can.*, 35, 630–636.

Goldberg, E.D., Bowen, V.T., Farrington, J.W., Harvey, G., Martin, J., Parker, P.L., Risebrough, R.W., Robertson, W., Schneider, E., and Gamble, E. 1978. The Mussel Watch, *Environ. Conser.*, 5, 101–125.

Goldberg, E.D., Koide, M., Hodge, V., Flegal, A.R., and Martin, J. 1983. U.S. Mussel Watch: 1977–1978 results on trace metals and radionuclides, *Estuarine Coastal Shelf Sci.*, 16, 69–93.

Grant, J. and Cranford, P.J. 1991. Carbon and nitrogen scope for growth as a function of diet in the sea scallop, *Placopecten magellanicus*, *J. Mar. Biol. Assoc. U.K.*, 71, 437–450.

Kramer, K.J.M., Jenner, H.A., and de Zwart, D. 1989. The valve movement response of mussels: a tool in biological monitoring, *Hydrobiologia*, 188/189, 433–443.

Martin, M. 1985. State Mussel Watch: toxics surveillance in California, *Mar. Pollut. Bull.*, 16, 140–146.

O'Connor, T.P., Cantillo, A.Y., and Lauenstein, G.G. 1994. Monitoring of temporal trends in chemical contamination by the NOAA National Status and Trends Mussels Watch project, in Kramer, K.J.M., Ed., *Biomonitoring of Coastal Waters and Estuaries*, CRC Press, Boca Raton, FL, 29–50.

Pitts, L.C. and Wallace, G.T. 1994. Lead deposition in the shell of the bivalve, *Mya arenaria*: an indicator of dissolved lead in seawater, *Estuarine Coastal Shelf Sci.*, 39, 93–104.

Rapport, D.J., Regier, H.A., and Hutchinson, T.C. 1985. Ecosystem behavior under stress, *Am. Nat.*, 125, 617–640.

Riisgård, H.U. and Randløv, A. 1981. Energy budgets, growth and filtration rates in *Mytilus edulis* at different algae concentrations, *Mar. Biol.*, 61, 227–234.

Roesijadi, G., Young, J.S., Drum, A.S., and Gurtisen, J.M. 1984. Behavior of trace metals in *Mytilus edulis* during a reciprocal transplant field experiment, *Mar. Ecol. Prog. Ser.*, 15, 155–170.

Smaal, A.C. and Widdows, J. 1994. The scope for growth of bivalves as an integrated response parameter in biological monitoring, in Kramer, K.J.M., Ed., *Biomonitoring of Coastal Waters and Estuaries*, CRC Press, Boca Raton, FL, 247–267.

Waldock, M.J. and Thain, J.E. 1983. Shell thickening in *Crassostrea gigas*: organotin antifouling or sediment induced?, *Mar. Pollut. Bull.*, 14, 411–415.

Warren, C.E. and Davis, G.E. 1967. Laboratory studies on the feeding bioenergetics and growth in fish, in Gerking, S.C., Ed., *The Biological Basis of Freshwater Fish Production*, Blackwell Scientific, Oxford, 175–214.

Widdows, J. 1991. Physiological ecology of mussel larvae, *Aquaculture*, 94, 147–164.

Widdows, J. and Donkin, P. 1992. Mussels and environmental contaminants: bioaccumulation and physiological aspects, in Gosling, E., Ed., *The Mussel Mytilus: Ecology, Physiology, Genetics and Culture*, Elsevier, Amsterdam, 383–424.

Widdows, J. and Johnson, D. 1988. Physiological energetics of *Mytilus edulis*: scope for growth, *Mar. Ecol. Prog. Ser.*, 46, 113–121.

SYNTHESIS and CONCLUSIONS

9 Conclusions

An ecosystem perspective following the process–functional school has been used in this book to bring together the different ecological aspects of marine bivalves. This approach was utilized for two principal reasons: (1) it incorporates the elements of organismic, population, community, and ecosystem ecology in order to provide a synthesis to the study of bivalve ecology; and (2) most of bivalve ecology to date has focused on organismic or population level analyses, and the ecosystem approach offers a new and different perspective to the role bivalves play in their respective systems. The population–community and process–functional approaches ask different types of questions that address ecological structure and function. In the older population and community school, structure refers to the distribution and abundance of organisms and the consequences of their interactions with each other and their environment. Studies at the ecosystem level, however, often focus on understanding the effects of ecosystem processes, interactions, and components on their context, or on the next larger organizational scale (Grimm 1995).

Ecosystem level studies are generally based on the theories or laws of biogeochemistry generated from elemental stoichiometry and energy flow derived from thermodynamics (Reiners 1986). In practice, the information needed for the ecosystem approach is often derived from the aggregation of fine-grained observations using organismic (Chapters 2 and 3), population, and community techniques (Chapter 4) and applying these data to functional groups, subsystems, and ecosystems (Chapters 5, 6, and 7). Incidental to these ecosystem oriented observations, information on organismic processes, species composition, and population dynamics is obtained (Grimm 1995).

In using the ecosystem approach, care should be taken to not lose the identity of the species that occupy a given functional group, i.e., the black box. The occupants of the group offer an indication of system level functional redundancy (Lawton and Brown 1993). In addition, they indicate which species may be keystones to the operation of the ecosystem (Grimm 1995).

KEYSTONE SPECIES

Paine (1969) first utilized the term "keystone" with regard to predators in the rocky intertidal zone. Since then, the idea has been extended to species whose removal from the system results in significant change in community structure or ecosystem functioning (Grimm 1995). For ecosystems, keystone species are those species whose contribution or role in ecosystem functioning is unique or essential. The functional role of these species can be loosely categorized into trophic, biogeochemical, and structural–functional groups. For marine ecosystems with significant bivalve components, the preceding chapters have examined numerous cases where bivalves can be considered keystone species in a given system, e.g., oyster reefs, mussel beds, cockle beds, clam beds, etc.

ECOSYSTEM ENGINEERS

In addition to the keystone species concept, Jones et al. (1994) and Lawton and Jones (1995) have used the term "ecosystem engineers" to describe species that directly or indirectly modulate the availability of resources to other species, by causing physical state changes in biotic or abiotic materials. In doing so, these engineering organisms modify, maintain, and create habitats. These authors further argue that it is a critical character of ecosystem engineers that they must change the quality, quantity, and distribution (availability) of resources used by other components, excluding the biomass provided directly by prey organisms. Thus, in building a reef or bed, changing the biogeochemistry of the sediments through deposition and bioturbation, maintaining benthic grazing versus planktonic grazing, bivalves in a given system change the environment and modulate the distribution and abundance of many resources. These engineering impacts are, as we have seen, much more obvious when the bivalves are living in high densities.

ECOSYSTEM LEVEL ROLES OF BIVALVES

To provide an overall ecosystem perspective, the various roles of bivalves in marine systems will be reviewed and references to details in pertinent citations and preceding chapters noted.

TROPHIC

Bivalves can be categorized into several feeding or trophic groups, and it is often through these feeding processes that large quantities of energy and matter enter their respective habitats. Suspension or filter feeding bivalves are very common in shallow coastal marine habitats with firm to rocky substrates (Jørgensen 1990). These animals commonly graze (Chapter 5) on phytoplankton in the water column, resuspended benthic microalgae, protozoans, and organic detritus with attached microflora (Smaal and Prins 1993). They may also indirectly prey upon their own larvae (Chapter 3). In systems with low water residence time, relatively low phytoplankton growth rates, and dense populations of bivalve filter feeders, phytoplankton may be regulated through grazing (Officer et al. 1982; Smaal and Prins 1993; Heip et al. 1995; and Chapter 5). In ecosystems with large bivalve filter feeder components, system stability is thought to be enhanced through their strong coupling to the water column and their relatively long life spans (Herman and Scholten 1990). The reduction or demise of the bivalve filter feeders in an ecosystem commonly dominated by them can lead to dramatic changes or phase shifts to water column-dominated food webs (Newell 1988; Chapter 5).

Deposit feeding bivalves are typically found on or in soft or muddy sediments where they consume sediments and remove organically rich particles (Chapter 3). In this deposit feeding mode, bivalves are in competition with the sediment microbial community for food, and they may also accelerate microbial activity and consequently the decomposition of organic matter (Lopez and Levinton 1987; Levinton 1995). It has been argued by Rhoads and Young (1970) that deposit feeders change the benthic environment so as to hinder the filter feeders. This action results in bivalve filter feeders in one area and bivalve deposit feeders in another. Thayer (1979) has carried this argument further, to an evolutionary scale where deposit feeders (Thayer refers to as biological bulldozers) may have caused the extinction of many benthic filter feeding groups. These arguments suggest a spatial phase shift and possibly a temporal phase shift between deposit feeders and filter feeders.

Symbiotic feeding relationships are found in a number of different bivalves. Photoautotrophic symbiosis is prevalent in the giant clams (Tridacnidae) common to Pacific coral reef systems. In these bivalves, feeding is partitioned between normal filter feeding, uptake of dissolved organic matter, uptake of photosynthates generated by symbiontic zooxanthellae living within the tissues of the clams, and digestion of zooxanthellae cells (Fankboner and Reid 1990; Klumpp et al. 1992; and Chapter 3). Zooxanthellae nutrition dominates in the adult clams and filter feeding dominates in the young animals. The symbiotic relationship allows these clams to be nutritional opportunists and in those species with rapid growth rates, to grow faster than the surrounding corals.

Chemoautotrophic symbiotic relationships are found in bivalves in a number of extreme environments, e.g., deep sea hydrothermal vents and oil seeps and anaerobic sediments (Fisher 1990; Chapter 3). In these symbioses, bacteria living in the bivalve's tissues convert inorganic chemicals, usually reduced sulfur, into organic compounds that can be utilized by the animals. Chemoautotrophic symbiosis allows bivalves to inhabit environments where they normally could not survive.

Nutrient Cycling

Bivalves process materials by consuming particulate and dissolved organic matter and excreting inorganic nutrients. In performing this process, they may couple the water column to the benthos, bioturbate the sediments, and change the biogeochemistry of the adjacent substrates (Dame 1993; Levinton 1995; Chapter 6). In essence, they can become major components in the cycling of nutrients with their ecosystems, particularly when they are abundant. Bivalves process materials at high rates and, thus, they speed up nutrient cycling. Because their bodies are relatively massive and their life spans are long when compared to plankton, the materials sequestered within their tissues can be retained within the ecosystem more efficiently than would be expected of plankton. Furthermore, nutrient processing by bivalves, both directly by the animals and indirectly in adjacent sediments, increases the concentrations of limiting nutrients within their environments and in doing so may regulate primary production.

Structural

Bivalves may influence their immediate environment by being part of or the major component of the structural matrix, e.g., coral reefs, bivalve beds, or sediment fabric

(Chapters 3 and 4). At this scale, the structure of their bodies and burrows in conjunction with their functional activities influences the biodiversity of their communities and serves as a focal point for predators. As noted above, all bivalves can affect the biogeochemistry of the adjacent sediments through nutrient processing, and these activities probably lead to major changes in the diversity of organisms in the sediments.

On a larger spatial scale, filter feeding bivalves through their grazing and nutrient regeneration processes can regulate the diversity of phytoplankton in the water column (Prins et al. 1995, Chapters 5, 6, and 7). Furthermore, strong benthic-pelagic coupling by bivalves may dictate one phase in a coastal system while loss of this coupling may institute a shift to a system dominated by planktonic food webs (Newell 1988).

The flows of matter and energy between bivalve food resources, predators, and nutrient pools connect these animals in a multitude of ways to other ecosystem components. This flow structure sets up both positive and negative feedback loops that provide additional diversity, stability and structure in their ecosystems (Ulanowicz 1986; Chapter 7).

Monitors and Indicators

Marine bivalves are proven indicators of stress in marine ecosystems (Chapter 8). Because they are sessile as adults and pump water through their bodies, they are recognized as good monitors of ecosystem change over time scales varying from hours to years. In addition, the chemistry of their shells can provide valuable information on global changes in the environment.

ECOSYSTEM ISSUES

In many marine ecosystems, bivalves can be considered important components worthy of the terms keystone species and/or ecosystem engineers. From the discussions in this book, clearly some areas of bivalve ecological research are much more developed than others. For example, studies of commercially important bivalve filter feeders dominate the literature, while studies of other types of bivalves are relatively few. In this section, I will highlight a few of the major questions and issues that need to be addressed to further our understanding of the roles of bivalves in marine ecosystems.

Organismic Level

While the difficulties surrounding the role of water flow in the feeding processes of filter feeding bivalves are gradually being understood, there is still a need for similar quality information on deposit feeders and symbiotic forms in order to better understand how these bivalves function in their respective ecosystems.

Food quality and how it influences bivalve growth and excretory products needs to be extensively investigated, particularly from the aspect of the basic elements of carbon, nitrogen, phosphorus, and silicon so as to elucidate the role of all trophic types of bivalves at the ecosystem level.

The role of bivalves feeding on components of the microbial loop has important implications on bivalve couplings to the water column and the sediments, but little is known of the importance of these processes.

Population Level

Because of their meroplanktonic larvae, bivalve populations may not behave as classical closed populations, but as open or metapopulations. The metapopulation aspect has only recently been addressed in filter feeders, but is equally important to the understanding of the spatial and temporal distribution of all types of bivalves.

Although many classical studies have been conducted on the predation of rocky intertidal bivalves, there are few quantitative works on predation of bivalves in other habitats. In particular, the role of spatial and temporal influences on predators foraging on bivalves is critical to understanding the role of bivalves in ecosystem processes.

The role of bivalves in structuring their respective communities has been only cursorily studied for a few common groups. There is little understanding of the observed temporal and spatial variability or how this may impact ecosystems.

Ecosystem Level

The contrast between deposit feeding and filter feeding bivalves was recognized many years ago, but little effort has been expended by bivalve ecologists, most of whom are filter feeder specialists, in examining the potentially competitive roles of these two groups. Their relationship may be an example of both a temporal and spatial phase shift between benthic systems that might be freshly examined with an ecosystem approach.

Another potential phase change in coastal systems that is worth investigating from an ecosystem perspective is the shift between benthic–pelagic bivalve filter feeder dominance and pelagic food web prevalence. This issue raises numerous pertinent questions related to ecosystems, including habitat change, biodiversity, functional redundancy, and overall nutrient processing.

There is sufficient existing information on many systems dominated by bivalve filter feeders to suggest that the commonalty among these systems can be represented by a generic model including parameters for water dynamics, phytoplankton, and the bivalves. Such an approach might also cast light on the benthic–pelagic food web phase shift question as well as system connectivity and ascendancy among systems.

The ecosystem role of both chemoautotrophic and photoautotrophic bivalves is at best speculative. The utilization of the ecosystem approach in these fundamentally different systems should be a fertile area of research.

The preceding issues are for the most part basic science and only incidentally address applied questions. However, in the area of ecosystem health, bivalves have already proven themselves useful monitors and indicators of ecosystem stress and change. It would seem appropriate that systems with dominant bivalve communities should be enlisted into efforts to develop measures of ecosystem health.

SYNTHESIS

Möbius (1880) noted that the organisms living on oyster beds were a positive influence on each other and also interacted with the surrounding nonliving environment. His statement is often cited as the first formal expression of a unity of organisms and their environment from whence the ecosystem school of thought eventually was derived. It has been more than a century since Möbius' study on oyster beds and ecology has developed in dramatic and sometimes unforeseen ways. Today ecosystem level studies are a formal and respected area of science that encompass the techniques and theories of other areas as well as its own. It is this synthetic approach that has been woven into this book.

Using marine bivalves as a focus, this work describes the ecology of bivalves from an organismic to an ecosystem perspective. The theories and techniques of many areas of science are used to support the various approaches. With these building blocks provided as a hierarchical framework, the reader should be able to conceptualize the role of marine bivalves in many different marine ecosystems. It is hoped that with this stimulus new questions will be posed, new research conducted, and our scientific understanding of the ecology of marine bivalves will be forwarded.

REFERENCES

Dame, R.F. 1993. The role of bivalve filter feeder material fluxes in estuarine ecosystems, in Dame, R.F., Ed., *Bivalve Filter Feeders in Estuarine and Coastal Ecosystem Processes*, Springer-Verlag, Heidelberg, 245–269.

Fankboner, P.V. and Reid, R.G.B. 1990. Nutrition in giant clams (Tridacnidae), in Morton, B., Ed., *The Bivalvia — Proc. Mem. Symp. in Honour of Sir Charles Maurice Yonge, Edinburgh, 1986*, Hong Kong University Press, Hong Kong, 195–209.

Fisher, C.R. 1990. Chemoautotrophic and methanotrophic symbioses in marine invertebrates, *Rev. Aquat. Sci.*, 2, 399–436.

Grimm, N.B. 1995. Why link species and ecosystems? A perspective from ecosystem ecology, in Jones, C.G. and Lawton, J.H., Eds., *Linking Species and Ecosystems*, Chapman and Hall, New York, 5–15.

Heip, C.H.R., Goosen, N.K., Herman, P.M.J., Kromkamp, J., Middelburg, J., and Soetaert, K. 1995. Production and consumption of biological particles in temperate tidal estuaries, *Annu. Rev. Oceanogr. Mar. Biol.*, 33, 1–149.

Herman, P.M.J. and Scholten, H. 1990. Can suspension-feeders stabilize estuarine ecosystems?, in Barnes, M. and Gibson, R., Eds., *Trophic Relationships in the Marine Environment*, Aberdeen University Press, Aberdeen, 104–116.

Jones, C.G., Lawton, J.H., and Shachak, M. 1994. Organisms as ecosystem engineers, *Oikos*, 69, 373–386.

Jørgensen, C.B. 1990. *Bivalve Filter Feeding: Hydrodynamics, Bioenergetics, Physiology, and Ecology*, Olsen and Olsen, Fredensborg, Denmark, 140 pp.

Klumpp, D.W., Bayne, B.L., and Hawkins, A.J.S. 1992. Nutrition of the giant clam *Tridacna gigas* (L.). I. Contribution of filter feeding and photosynthesis to respiration and growth, *J. Exp. Mar. Biol. Ecol.*, 155, 105–122.

Lawton, J.H. and Brown, V.K. 1993. Redundancy in ecosystems, in Schulze, E.D. and Mooney, H.A., Eds., *Biodiversity and Ecosystem Function*, Springer-Verlag, Berlin, 255–270.

Lawton, J.H. and Jones, C.G. 1995. Linking species and ecosystems: organisms as ecosystem engineers, in Jones, C.G. and Lawton, J.H., Eds., *Linking Species and Ecosystems*, Chapman and Hall, New York, 141–150.

Levinton, J.S. 1995. Bioturbators as ecosystem engineers: control of the sediment fabric, inter-individual interactions, and material fluxes, in Jones, C.G. and Lawton, J.H., Eds., *Linking Species and Ecosystems*, Chapman and Hall, New York, 29–36.

Lopez, G.R. and Levinton, J.S. 1987. Ecology of deposit-feeding animals in marine sediments, *Q. Rev. Biol.*, 62, 235–260.

Möbius, K. 1880. Die Auster und die Austerwirthschaft, *Rep. U.S. Comm. Fish.*, 683–751.

Newell, R.I.E. 1988. Ecological changes in Chesapeake Bay: are they the result of overharvesting the American oyster, *Crassostrea virginica*?, in Lynch, M.P. and Krome, E.C., Eds., *Understanding the Estuary: Advances in Chesapeake Bay Research*, Chesapeake Research Consortium, Solomon's, Maryland, 536–546.

Officer, C.B., Smayda, T.J., and Mann, R. 1982. Benthic filter feeding: a natural eutrophication control, *Mar. Ecol. Prog. Ser.*, 9, 203–210.

Paine, R.T. 1969. A note on trophic complexity and community stability, *Am. Nat.*, 103, 91–93.

Prins, T.C., Escaravage, V., Smaal, A.C., and Peeters, J.C.H. 1995. Nutrient cycling and phytoplankton dynamics in relation to mussel grazing in a mesocosm experiment, *Ophelia*, 41, 289–315.

Reiners, W.A. 1986. Complementary models for ecosystems, *Am. Nat.*, 127, 59–73.

Rhoads, D.C. and Young, D.K. 1970. The influence of deposit-feeding organisms on sediment stability and community trophic structure, *J. Mar. Res.*, 28, 150–178.

Thayer, C.W. 1979. Biological bulldozers and the evolution of marine benthic communities, *Science*, 203, 458–461.

Ulanowicz, R.E. 1986. *Growth and Development: Ecosystems Phenomenology*, Springer-Verlag, New York, 203 pp.

INDEX

INDEX

C

D

E

F

G

H

I

J

K

L

M

N

O

P

R

T

U

V